THE DYNAMICS OF DISASTER

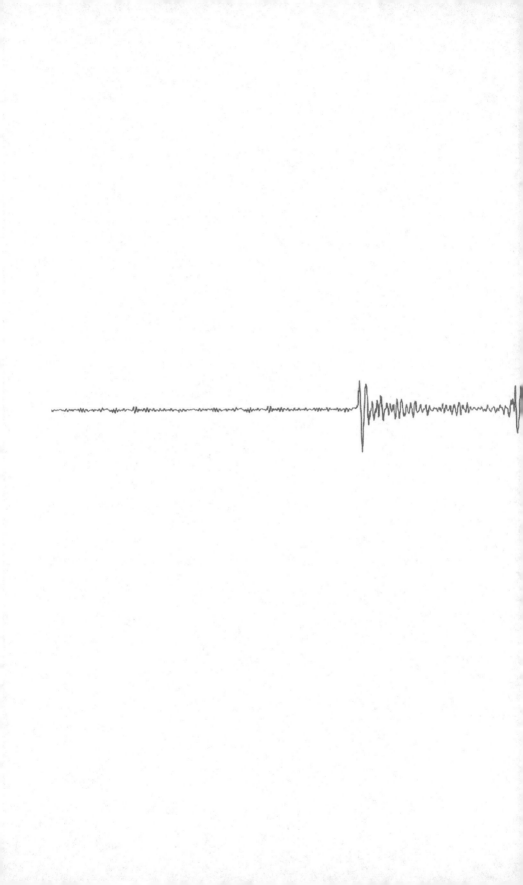

THE DYNAMICS
OF DISASTER

SUSAN W. KIEFFER

 W. W. NORTON & COMPANY New York London

For information about permission to reproduce selections from this book,
write to Permissions, W. W. Norton & Company, Inc.,
500 Fifth Avenue, New York, NY 10110

For information about special discounts for bulk purchases, please contact
W. W. Norton Special Sales at specialsales@wwnorton.com or 800-233-4830

Manufacturing by Courier Westford
Book design by Lovedog Studio
Production manager: Devon Zahn

Library of Congress Cataloging-in-Publication Data

Kieffer, Susan W.
The dynamics of disaster / Susan W. Keiffer. — First ed.
p. cm.
Includes bibliographical references and index.
ISBN 978-0-393-08095-7 (hardcover)
1. Natural disasters. I. Title.
GB5014.K54 2013
363.34—dc23
2013018412

W. W. Norton & Company, Inc.
500 Fifth Avenue, New York, N.Y. 10110
www.wwnorton.com

W. W. Norton & Company Ltd.
Castle House, 75/76 Wells Street, London W1T 3QT

1 2 3 4 5 6 7 8 9 0

With deepest gratitude to my "wise men"—E-an, Paul, Pete, and Ward—for decades of intellectual nourishment

and

to my wonderful, supportive husband and fellow adventurer, Jerry, who wondered why I was so fascinated that bananas were US$6 per pound in Australia!
(Reader: the answer is in Chapter 9!)

CONTENTS

THE NATURE
OF DISASTER

THE WORD "DISASTER" MEANS "A SUDDEN CALAMI-
tous event bringing great damage, loss, or destruction."[1] The
word could have meaning in the absence of humans; for example,
it could be used to describe the effect of a volcanic eruption on the
atmosphere. However, without a human to judge whether that is
a disaster for the atmosphere or not, that concept remains in the
world of academic discourse.

Most commonly, the word "disaster" is used to describe losses
and damages that humans experience, individually or collectively.
It is a word that conjures up a collage of images, ranging from the
seemingly calamitous loss of a first youthful love, to the profoundly
devastating loss of hundreds to even hundreds of thousands of people in violent acts of nature or humans. The singular word "disaster" reflects a spectrum of individual disasters ranging from purely
geological, such as a volcanic eruption, to purely human, such as
purposeful or accidental detonation of a nuclear device. This book
is about one form of disaster—that caused by the ongoing geological processes on our planet—and the intersection of the processes
that cause disaster with our human presence on the planet. In other
words, this book is about the dynamics of disaster.

The geological processes of our planet affect us in good ways, by providing us the essentials for life, and in bad ways—in disasters. In turn, we affect parts of Earth as well because all of its components are interconnected. In discussing disasters with a wide variety of people, I have found it useful to define two end members of the spectrum of disasters: "natural disasters" and "stealth disasters." Natural disasters result from the ongoing geological processes on our planet and have typically been referred to as "acts of God" by the insurance industry. They often have a sudden onset and immediate consequences, such as an earthquake, hurricane, or volcanic eruption. Some natural disasters, such as those that are weather related, are increasingly affected by humans.

Stealth disasters, in contrast, are caused by humans but involve the natural systems that support us. They have a gradual onset but near-term consequences. Some examples of stealth disasters are climate change, desertification, loss of soils, collapse of aquifers that store water underground, and acidification of the oceans. Climate change is the best-publicized stealth disaster these days. While much has been written about the causes of climate change, including the impact of natural processes such as volcanic eruptions on climate, much less has been written about how climate change—or any other stealth disaster—is likely to influence natural disasters. The notable exception is the expected impact of climate change on atmospheric disturbances, such as number and intensity of hurricanes. But stealth disasters, like climate change, can—and are likely to—influence other natural disasters. For example, the ongoing removal of ice from our polar regions that started at the end of the last ice age may affect the frequency and location of volcanic eruptions, the number and size of landslides, the operation of hot-water geothermal systems, the magnitude and frequency of river floods, and even the frequency and style of earthquakes.

If we are to understand possible future effects of humans on the planet and on natural disasters, we need to understand the basic geological dynamics of natural disasters. In the past, our general perception of Earth has been that natural disasters are acts of nature that affect us. Now, however, we need also to understand how human behavior might impact natural disasters. Within such a framework the impact of stealth events needs to be broadened beyond climate change to all disasters, not just weather-related events. For example, how does climate change influence not only the weather, but the frequency of landslides, earthquakes, and volcanic eruptions? How does desertification affect not only regional weather and climate in Africa, but the frequency and strength of hurricanes elsewhere around the planet? How do such changes affect other natural processes, such as earthquakes and volcanic eruptions, at great distances?

The framework presented in this book is that natural disasters result from changes of state in a system through modifications of its materials, energy, or both, and that by understanding what causes these changes of state, we can begin to understand the possible influence of stealth disasters on natural disasters. Changes of state arise in several different ways. The two most common ways are by modifications of the materials in a system, and by modifications of their conditions of motion. An earthquake might rip open the earth, sending out vibrations that shake the ground, turning soils to mush, triggering landslides, or generating tsunamis.[2] A river might flood, causing destructive waves or washing away its own banks. Winds might perturb the state of the ocean and atmosphere, causing rogue waves that sink ships in the oceans or atmospheric waves that attack airplanes in the atmosphere. Changes of pressure in the atmosphere, in the weight of glaciers on Earth's surface, and in temperature or the migration of fluids in Earth's crust can cause instabilities that trigger geothermal or volcanic eruptions.

Once the change of state involved with a particular natural disaster is understood, we can begin to institute engineering and policy practices to minimize its impact on our lives and communities. How do we deal with common natural disasters, such as small or moderate earthquakes? How do we deal with bigger events that are rare and have a low probability of occurrence, but have very high human and economic consequences? Do we deal with these two cases in the same way? How do we balance our attention to natural disasters with attention to other issues in our personal and communal lives—poverty, disease, hunger? And, perhaps most importantly for the scope of this book, how are natural disasters in the future going to compare with those of the past in impacting our lives on this ever-more-crowded planet? Is it possible that our sheer numbers will exacerbate the type and magnitude of natural disasters?

A book could be written on each of the preceding questions alone. I examine some of these questions in this book but there is not enough space for detailed answers. Rather, this book provides the basic science needed to answer the questions. The issues are a subject of deep concern among geologists at present, and they will remain for future generations, and for a community far larger than one person or, indeed (as I reflect in Chapter 10), for the scientific community alone, to address. They are for the global community of scientists, engineers, political and spiritual leaders, and citizens. It is my hope that this book, and its concluding chapter calling for a CDC-PE (a Center for Disaster Control for Planet Earth), will challenge readers to explore the interconnectedness of everything on this planet and to think about the health of the planet as a whole and our role as humans in keeping the planet healthy for present and future generations.

Chapter 1

GEOLOGIC CONSENT— DO WE HAVE IT OR NOT?

> Civilization exists by geologic consent, subject to change without notice.[1]
>
> —*Historian Will Durant*

CHANGES OF STATE AND CHANGE
WITHOUT NOTICE

Contrary to popular belief, but confirmed by repeated natural disasters that have taught us otherwise, humans have amazingly little control over Mother Nature. Yet we persist in building centers of our civilization in places of known past disasters, and when they are destroyed we rebuild in the same places, always believing that our wits and technology will do better the next time around (Figure 1.1). Rarely do we win these battles with Earth. It may take one, hundreds, or even thousands of years, but Mother Nature always strikes again.

Nature's destructive events are so diverse that it is difficult to grasp the variety of processes the planet is hurling at us. Some of us live with, and understand, earthquakes. Others coexist with floods, and still others with landslides, tornadoes, or tsunamis. Sometimes we fear these events, but strangely, at other times we are fascinated by them—the beauty of volcanoes, the sport of rafting on big rivers, the challenges of sailing big ships in danger-

FIGURE 1.1 Mount Vesuvius, in the center of this image, is surrounded by the densely populated city of Naples, Italy, where 3 million people live on deposits from the massive eruption in AD 79. *Image courtesy of NASA/ GSFC/MITI/ERSDAC/JAROS and US/Japan ASTER science team.*

ous waters or of setting world records for balloon ascents. And certain of our endeavors—sailing merchant or military ships on the oceans, or drilling deep into reservoirs to extract resources, for example—force us to face the extremes of nature as challenges head-on.

I am a planetary scientist, a scientist studying other worlds through the eyes of a geologist. My academic life has focused on the science of powerful geological forces—volcanic eruptions, river floods, and meteorite impacts. I have worked with the science of these phenomena so long that even geological events big enough to make the news don't surprise me very much. Hurricane Katrina? Probably overdue, could have been worse, and will happen again on the Gulf Coast. Mount St. Helens eruption of 1980? Puny. The 2011 earthquake on the East Coast? Punier, but definitely interesting. The 1991 Pinatubo eruption in the Philippines? Definitely worth attention, but expect another similar one somewhere within a decade or two. The magnitude 9 Tohoku earthquake and tsunami of 2011? I'm sure many wish Earth had gotten that one out of her system earlier! Unfortunately, the global average for magnitude 9 earthquakes is about one every thirty years.

In the course of giving talks on the dynamics of natural disasters to people who live in regions with different hazards, it has gradually dawned on me that there are real knowledge and communication gaps. People living with, and knowledgeable about, one kind of disaster rarely have much knowledge (or even interest sometimes) in other types of disasters. The people who have the most knowledge about disasters in general are in the insurance industry. And even they sometimes do not have the right knowledge. The insurance industry is based on actuarial statistics, which is fine for events that occur frequently, like annual spring floods. But it is a flawed analytical tool when applied to big and rare events. The momentous flood that statistically occurs only every hundred years could occur tomorrow, and it could happen again next year as well. Or it may not happen for a few hundred years. The fact that, on average, there are three magnitude 9 earthquakes each century did not prevent there being five in the last half century. The global insurance industry was battered by disasters in the first decade of the twenty-first century, and as a result, insurance companies are realizing more and more that they need information on the fundamental science underlying the disasters that they are insuring against.

All this set me to wondering, do our political leaders and their advisers have an overview about the number and magnitude of possible disasters? Sadly, it seems that the big picture is too often lacking at these levels. In September 2011, the Federal Emergency Management Agency (FEMA) ran out of money after fewer than nine months of hurricanes, tornadoes, wildfires, and droughts across the US.[2] In just the first nine months of the year, bills were running at a trillion dollars, and Congress was deadlocked on how, or even whether, to grant the agency more money. Bridges, roads, schools, and homes damaged in these events could not be repaired. Debate raged about the relative responsibilities of insurance companies and local, state, and federal governments.

The situation is not new; much damage on the Gulf Coast of the US from Hurricane Katrina in 2005 has not yet been repaired—years after the tragedy. In 2011, FEMA officials complained not only that there had been more than the usual number of major disasters, but that over the years, more and more smaller events have been declared as qualified for relief funding, deflecting the agency from its primary mission of responding to truly large-scale events. In the words of former FEMA director James Lee Witt, disaster relief has become a "game"[3]—with different members of Congress using techniques to force FEMA funding to their own states even when the disasters do not meet FEMA's criteria for receiving aid.

As quoted at the start of this chapter, the historian Will Durant once said, "Civilization exists by geologic consent, subject to change without notice." On this heavily crowded, if not over-crowded, planet, there is no time for games, because the next disaster—to me, or my neighbor, or a total stranger, but still a fellow member of the human race—lies just around the corner. Somewhere, someone will need resources, and somewhere, someone else will be making decisions on how much is available and who gets it. Wise decisions can be made only in the broad context of the variety of disasters and how they occur. If we lived in a world with only one type of disaster—say, earthquakes—then it would be sufficient to understand and remediate against that single disaster. However, we live in a world in which disasters appear to be nearly infinitely varied. The task of preparing ourselves for every eventuality is daunting. How do we evaluate relative risks or equate risk from very different types of disasters?

There is an underlying unity in the dynamics of geological disasters—that they all result from changes in the distribution of energy within the earth, what I call "changes of state." These changes of state are the cause of the "change without notice" that Durant refers to. Those who understand this concept well will be best prepared to plan for, and to prevent or recover from, disasters.

DISASTERS: NATURAL, UNNATURAL, TECHNOLOGICAL, STEALTH

The roots of the word "disaster" date back to Middle French and Old Italian in the sixteenth century, when a word was needed to indicate a natural event believed to result from a bad alignment of planets or stars (the "astro" in "dis*aster*"). In modern times, and in official actions such as the declaration of a disaster area, a disaster is an event that causes great damage, loss, or destruction to humans. No longer believing that disasters result from unfortunate alignments of the planets, we now know that natural geological processes of our planet produce events that we view as disasters; we call these "natural disasters." Almost all natural disasters result from storms (hurricanes, cyclones, floods), earthquakes (shaking, liquefaction, tsunamis, landslides), and volcanic eruptions. We ourselves can also produce disasters of a magnitude rivaling natural disasters; these are human-caused, or "technological," disasters. The most common technological disasters of concern result from nuclear accidents, bombings, bioterrorism, and extreme drilling. In our densely populated, highly interconnected world of the twenty-first century, disasters fall along a spectrum blending natural and technological causes.

In the context of the insurance industry, the term "natural disaster" applies to the classic "acts of God" that are nonhuman in origin. These typically, but not always, have an abrupt onset, cause immediate and major change to the surface of the earth, and, in our current world of swift communication, get much media and public attention.[4] By contrast, some disasters have a protracted onset, and their consequences may or may not be obvious initially. Some of these slow-onset phenomena are clearly related to human activities, such as chronic loss of soil fertility, acidification and eutrophication of the oceans, deforestation, contamination of potable freshwater, and loss of biodiversity. Others, such as climate change, droughts, desertification, plagues, famines, pandemics, heat waves, and invasive species, appear to be a combination of natural and human causes, with the degree to which each cause contributes to the problem sometimes being highly controversial. For clarity, I have dubbed all slow-onset disasters as stealth disasters,[5] and they are beyond the scope of this book. Disasters span a range from purely natural to purely stealth.

The World Bank published a report with the intriguing title *Natural Hazards, UnNatural Disasters*,[6] in which disasters are classified as either "natural hazards" (the so-called acts of God) or "unnatural disasters" (disasters within disasters). (The discussion applies equally well to technological disasters, though they were not considered in this report.) In unnatural disasters, a primary hazard allows a second, hidden disaster—the disaster within the disaster—to unfold. Unnatural disasters are caused by acts and policies by individuals and governments that result in more deaths than a natural disaster by itself would cause. The extra deaths occur because cumulative results of many earlier decisions on many different scales have led to inadequate, or wrong, policies. Both Hurricane Katrina in the US and the Tohoku earthquake in Japan contain tragic examples of unnatural disasters.

This same report, published in 2010 before the Tohoku earthquake and tsunami, estimated that annual losses from natural disasters[7] would be as high as $185 billion by the end of the century, not counting losses due to climate change that could add $54 billion more per year. With the cost of the Tohoku events alone exceeding $200 billion, it is clear that these estimates were already much too low. Technological human-generated disasters, such as chemical/nuclear spills, dam failures, explosions, fires, and warfare, are increasingly of the magnitude, or possible magnitude, of natural disasters. Tragically, natural disasters and human-caused disasters are more and more intertwined as the population of the planet increases.

KNOWN KNOWNS, KNOWN UNKNOWNS, AND UNKNOWN UNKNOWNS

Our science regarding natural disasters and their causes is riddled with incompleteness and uncertainties. Almost by definition, disasters are events that happen infrequently and are thus difficult to study. Our knowledge about such events will always be incomplete. US Defense Secretary Donald Rumsfeld addressed the topic of incomplete knowledge in a widely publicized 2002 press conference[8] in a different context—the increasingly unstable military and political situation in Afghanistan at that time:

Reports that say that something hasn't happened are always interesting to me, because as we know, there are known knowns; *there are things we* know we know. We *also* know there are known unknowns; *that is to say we know there are some things we* do not know. *But there are also* unknown unknowns—*the ones we* don't know we don't know. *(emphasis added)*

Some ridiculed Rumsfeld, giving him the 2003 "Foot in Mouth Award" of the British company Plain English Campaign, for "a baffling comment by a public figure."[9] Others, including many scientists and philosophers, recognized that this statement invokes an important and ancient concept that dates as far back as Socrates: "I know as a non-knowing" or "I know that I don't know."[10] Awareness, unawareness, knowns, and unknowns are important topics covered by a large literature in economics, military theory, law, and science. They are crucial issues for optimizing our preparedness for disasters on this active planet that we call home.

Rumsfeld's statement can be applied to two frameworks: a broad framework of the state of collective knowledge, and the individual framework of personal knowledge. What do we collectively know? What small fragment of that collective knowledge do I know? How can I extract knowledge from the collective wisdom to reduce the number of unknown unknowns in my life? How can I insert what I know into the collective framework so that it may become useful to others?

In our individual lives, the "known knowns" are those elements of knowledge that are common to almost all humans.[11] Some known knowns are the patterns of day and night, the longer-term rhythms of seasons, the intricate relations of a family, basic interactions between humans, common rituals of existence, and the day-to-day processes in the world around us.

In detail, we do not all have the same exact known knowns, because the planet is so large and humans are so diverse. Some of us live in deserts where the behavior of blowing sand is a known known. Others live near rivers and understand their ebb and flow. Most people function almost intuitively about many known knowns of their normal world. We walk, cycle, or drive a car through our world of known knowns. In simple situations we function as independent individuals because we have enough knowledge of the world to do so. We take shelter from sandstorms or protect ourselves against drowning.

As more complex situations arise, we find a growing discrepancy between the state of collective knowledge and the state of individual knowledge. There are the infrequent events that we may not have experienced personally but know about via the collective knowledge passed down through human history or, these days, from television or the Internet. To name a few: tornadoes or cyclones that whip the normally calm atmosphere into a frenzy, sandstorms that block out the sunlight for days at a time, or the magnitude 9 earthquake that occurs only every few decades. We know from our collective consciousness that "terra" isn't so "firma."

Some events, however, happen so infrequently that there is great uncertainty when they will happen next and what they will be like. These are the "known unknowns." As a society, we have developed a few ways to address these—for example, with public policy, disaster preparedness, and insurance. Indeed, the insurance industry was spurred along specifically to deal with known unknowns, first by the Great Fire of London that destroyed more than 13,000 buildings in 1666, and then by the Lisbon earthquake of 1755.[12]

A tragic illustration of the magnitude of the known unknowns was the great Tohoku earthquake in 2011. We knew that great earthquakes occur, but the uncertainties in our knowledge prevented us from being able to say how, when, and where a catastrophe might happen, and how big that catastrophe might be. These uncertainties, in turn, blinded scientists to the fact that such an earthquake might cause a tsunami that would decimate northeastern Honshu.

The prevailing paradigm of the twentieth century was that the magnitude of an earthquake depended on the length of the fault that ruptured, so scientists look for evidence of older faults in Earth's crust. Small, short faults should produce only small to moderate earthquakes, and only large faults were thought to be able to produce large earthquakes. The great Sumatra earthquake of 2004, with an estimated magnitude of 9.1 to 9.3 (use 9.2 as an average), appeared to fit that model; it was caused by a rupture along an enormously long fault, about 800 miles. In Japan, such a large fault had been recognized southwest of Tokyo, and so great was the scientific consensus about the correlation between large faults and large earthquakes that the Japanese Parliament passed a law in 1978 making earthquake preparedness a priority in that southwest region.

Tragically, this paradigm was wrong. The fault that ruptured and caused the great magnitude 9 Tohoku earthquake was only about 125 miles long and was northeast, not southwest, of Tokyo. How could such a wimpy fault cause such devastation? The progression of science will be to build on what we have learned from this tragic event in hopes of doing better next time, but progress is slow and, all too often, expensive.

What are the "unknown unknowns"? I'm not trying to be deeply philosophical here, but rather to bring the incomplete state of our knowledge to your attention. Could the state of our existence—as we know it—possibly change dramatically in ways that we have not yet experienced or even imagined? In other words, is it possible that unknown unknowns are lurking? Remember Durant's quote: "Civilization exists by geologic consent, subject to change without notice." We could take this to mean that there is no notice, no precursor, to a geological event. In another sense, we might consider the extremely rare event to be an unknown unknown, remembering that modern human history with good records of historical events is amazingly short compared to geological time. In this rather loose and vague sense, the Tohoku earthquake and tsunami event comes close to being an unknown unknown. In yet a third sense, we might more broadly think that we have no knowledge even of what is lurking, waiting to be unleashed somewhere, sometime, by our Earth. In this sense, these are geological black swans à la Nassim Taleb.[13]

The progression of human evolution, and in particular of science, has been to change unknown unknowns into known knowns (facts) with some associated known unknowns (the uncertainties associated with those facts). Two examples are the discoveries of plate tectonics and of extinction-causing meteorite impacts during the second half of the twentieth century. Realizing that these two processes, which seem so obvious to us now, were unknown only a century ago causes many scientists to be haunted by the possibility that there may be changes of state in store for our planet that have yet to be discovered—the unknown unknowns. This is where geologists can bring a unique perspective to bear: we can look through the long stretches of geological time to try to find, and understand, what that record tells us about processes on Earth. The role of geologists like me is to unravel the possible magnitude of these events and bring that understanding to bear on processes in our current environment. In this sense, we are trying to turn geological black swans white.

Once these processes are brought into the realm of known knowns with their associated uncertainties, people need to be made aware of the processes on the earth around them, including disasters. The need for education about disasters and their causes was the first policy implication pointed out in the World Bank report cited earlier. In spite of the complexities of civilized societies, humans and our social constructs remain remarkably unprepared for unexpected changes of state, particularly those that are relatively rare.[14] Preparing for the rare event is expensive, and it seems not urgent compared to more pressing and immediate problems. Yet understanding how changes of state occur, and what consequences they have on a geological scale, is critical both to our personal survival and to the survival of civilized societies. We may not be able to prevent disasters, but we can prepare for and remediate them. Such efforts require that the private and public sectors, individuals, and governments work well together. Such cooperation, in turn, requires that individual citizens understand, and modify, dangerous practices and behavior. By understanding the dynamics of disasters and viewing them within the context of changes of state, we can see the underlying unity of causes, apply lessons from one disaster to others, and ultimately arrive at personal and collective decisions and policies to minimize the effects of natural disasters in our lives.

A TOUR OF DISASTERS

What drives natural disasters? If we understand the underlying dynamics, can we control them? Or at least mitigate their effects? Or will we remain relatively powerless in the face of them? What can we do to live in harmony with our planet? Is it even possible? Or are humans always going to be at odds with the nonhuman world? To address these and other questions, let me lead you on a tour that will take us around the world and include investigations into the dynamic forces and changes of state that drive disasters.

I think of this book and its chapters as a bit like a typical geological field trip. One way that geologists from different parts of the world build a common body of knowledge is by visiting each other's areas of research, often in groups at so-called field conferences. Such conferences are convened not in the giant convention centers of cities, but at small and often remote sites where the rocks or topics under consideration are visible and interesting. A conference often starts with a series of lectures that give an overview of the problems, current research, and research questions, and a preview of the relevance of sites that will be visited. Participants then go out and wander around the rocks, bang on them with hammers, or look at them with a hand lens. There is ample, and often lively, discussion about how the rocks resemble, or differ from, rocks that other geologists around the world have seen. Then, typically, we reconvene at night in a comfortable setting with libations in hand, to talk about details or the big picture.

My geological field trip around the world is intended to give you a sense of the remarkable diversity in the dynamics of Earth and its disasters, with a focus on processes that occur on timescales of decades to centuries, occasionally millennia, but not eons.[15] We need to look around the world because, just as geologists studying rocks in the isolation of their own field area see only the local rocks and need to travel to learn about the remarkable diversity of rocks in other places in the world, we individually can learn only local lessons about disasters from our immediate environment. If we are exposed to only one type of disaster by our local environment, we can't get a perspective on disasters elsewhere. Fortunately, the modern global connectivity means that we have access to information about the state of the planet, so one of my goals is to motivate you to take advantage of this access to broaden your own perspective. Without this spatial perspective, self-interest will rule, as appears to be the case when members of Congress argue about how to divide up the FEMA pot while innocent victims of disasters wait months, or even years, for relief.

We need to look back through time as well, because the fact that human life is short means we are restricted in our experience in another critical way: we individually experience only a small sample of the ongoing processes on our planet—even if we're tuned in to the Internet. That is, in addition to spatial limitations on our perspectives, there are temporal limitations. Some disastrous processes occur only every few thousand or tens of thousands of years—for example, big volcanic eruptions. A geologist's view, however, is like a bifocal lens.[16] Geologists see the world at local to global scales, and although our lives are brief specks in time, we spend that time looking at both the present and the past, through time all the way back to the beginning of Earth more than 4 billion years ago. My goal is to share this bifocal view with you.

In the spirit of a field conference, before we begin our tour in this book I provide an introduction to the dynamics of disasters and their root causes—changes of state—in Chapter 2. Then we begin our field trips to various disasters. First, in Chapter 3, we'll go deep underground to see what happens when faults slip in earthquakes. Then we'll zip back up to the surface to look at one culprit that causes so much destruction during quakes: a process known as liquefaction. In Chapter 4 we'll look at landslides by taking a ride on the "flying carpet of Elm," a landslide that may have floated on a pocket of air before it destroyed a village in Switzerland. We'll examine the largest known landslide on Earth and discover that it shows evidence of intriguing processes just like some that occur during an earthquake. Landslides on the edifices of active volcanoes can initiate eruptions, and we'll visit Mount St. Helens in Washington State and Pinatubo in the Philippines to explore the dynamics of eruptions in Chapter 5.

Both landslides and volcanic eruptions can produce tsunamis, the topic of Chapter 6, if they occur near or in the oceans. Tsunamis are large when they approach land, but relatively benign and almost imperceptible on the open ocean. Large waves on the open ocean are not tsunamis, but rather wind-driven waves, and we will explore their generation and impact by visiting some of the most dangerous places in the ocean in Chapter 7. In Chapter 8 we'll go airborne to look at giant rivers of air that whip oceans into a frenzy and, over land, spawn deadly tornadoes. In Chapter 9, we'll visit Australia, where we see a continent that hosts alternately deadly droughts and deadly floods, and look at the relation of these conditions to El Niño or La Niña.

In each of Chapters 3 through 9, I have added a short section at the end entitled "Reflections," where I ponder, as geologists do on field trip evenings, issues broadly related to the specific topic of the chapter. In Chapter 10, I then conclude the book with an examination of the question "Do we have geologic consent for our civilization?" In this chapter I frame the elements that I believe would form the foundation for living with our active planet as a call for the creation—at least conceptually—of a world body loosely based on the model of the Centers for Disease Control and Prevention (CDC) of the US. Given the difficulty with interagency cooperation that we have seen in the US after Hurricanes Katrina and Sandy, this is probably a pipe dream, but the thought experiment shows how, if we have the collective will, we might reduce the impact of natural disasters on us, as well as our impact on the resources of the planet and its nonhuman inhabitants.

Chapter 2

DYNAMICS
AND
DISASTERS

EARTHQUAKES, TSUNAMIS, VOLCANOES

Before we take off on our field trips, let's look at the big picture and ponder the nature of disasters in general. Who could have predicted that in the first eleven years of the twenty-first century Mother Nature would unleash the costliest natural disaster in the history of the US (Hurricane Katrina, 2005), one earthquake and tsunami that would kill more than 300,000 people (Sumatra, 2004), another that would kill several hundred thousand (Haiti, 2010), a third that would kill tens of thousands, perhaps more (Szechuan Province in China, 2008), a rather small volcanic eruption that would disrupt global air traffic to the tune of $5 billion (Iceland's Eyjafjalljajökull, 2010), and, to top it off, an earthquake and tsunami that would devastate the northern coast of Japan and unleash a nuclear hell onto its people as the world watched (Tohoku, 2011)? Certainly not the insurance companies, or they would have raised their rates by a whole lot before these events occurred! All told, in these disasters plus others only slightly smaller, nearly a million people died. Damages will eventually total hundreds of billions of dollars. The US alone had at least fourteen disasters in 2011 that cost a billion dollars each.[1]

Disasters are dynamic. Usually, geological processes in Earth occur at, well, a geological pace! The weather is good, the seasons come and go, the ground is solid, and the oceans aren't in the news. But because Earth is a dynamic system, energy is constantly being stored in its parts in one form or another—as stress and strain in the crust, as heat in the atmosphere and oceans, or as energy in dissolved gases in the magma chambers that feed our active volcanoes. On a smaller and more personal scale, we have energy stored in various forms around us: heat is constantly being added to our water heaters, air is compressed in our automobile or bicycle tires, stress and strain are stored in the expanding tire that holds in the air, and chemical energy is available to us through the gasoline in our automobiles or natural gas that we use to heat our homes.

Disasters occur when accumulated energy is suddenly unleashed in a way that harms humans. It is a disaster when your water heater suddenly develops a crack and hot water and steam are released into your basement. It is a disaster when the car tire that stores energy in compressed gas blows out on a freeway. It is a disaster when the chemical energy in gasoline that powers our autos suddenly explodes and incinerates a car instead of powering it along the highway. In nature, disaster strikes in the form of storms, earthquakes, tsunamis, and volcanic eruptions when energy from the earth is unleashed upon us. During these events the normal state of the earth changes, usually briefly, to a different state—one that is dynamic, violent, and destructive. The release is episodic and often unpredictable. Let's look at some examples.

On January 21, 2010, an earthquake of Richter magnitude 7.0, centered only 16 miles west of Port-au-Prince, the densely popu-

FIGURE 2.1 Damage in Haiti due to the 2010 earthquake. *Photo by Walter Mooney, USGS.*

lated capital of Haiti, rocked the city and surrounding region.
Here, in one of the poorest nations of the world, people live in
shantytowns of flimsy construction. Their dwellings are constructed to shelter them from the tropical environment of heat,
winds, and drenching rains, but not to shield them from the
rare, major earthquake. Building codes are either nonexistent or
poorly enforced, but in the normal state of things these humble
dwellings provide shelter and respite from the elements to their
occupants because they are built on solid ground, terra firma.

However, the normal solid state of the ground changed dramatically during the 2010 earthquake. In many places the solid
ground shook violently enough to cause collapse of the flimsy
shantytown dwellings (Figure 2.1). In some other places, particularly near the port where buildings had been erected on artificial landfill, the solid ground turned to a mushy liquid-like state
(think of quicksand) that provided no supporting strength for the

structures. Although estimates vary widely, more than 50,000, and possibly several hundred thousand, people died in the earthquake, mostly victims of the poor infrastructure. Hundreds of thousands more were injured, and a million were made homeless.

Eight months later, on September 4, 2010, an earthquake of nearly the same magnitude (7.1) struck New Zealand only 15 miles from Christchurch, a city of nearly 400,000 people. Because the geological conditions were different, this earthquake produced much more severe ground vibration than in Haiti. Again, solid earth turned to mush, but in this case no one was killed, even though the infrastructure suffered major damage.[2] Here earthquake hazards were known, strict building codes were in existence, and those codes were enforced. These factors explain why there was such a difference in devastation between Haiti and Christchurch: the possibility that the normal state of the earth might change from solid to mush was largely ignored by one culture but was incorporated into the planning of the other.

As in New Zealand, the normal state of living in the coastal cities of Japan is one of preparedness for earthquakes and tsunamis because the Japanese people have a keen awareness of the long history of disasters that their tectonically and volcanically active island has visited on them. Buildings must meet strong code requirements, and many communities have built walls—some are 30 feet high—to protect them from storm surges and tsunamis. But even the hazard-conscious Japanese were unprepared for the magnitude of disaster that was unleashed on them in March of 2011. With no warning, a magnitude 9.0 earthquake, one of the strongest in recorded history, ripped apart the crust of the earth only 43 miles east of Honshu, the most populated island of Japan.

Now known as the Tohoku earthquake, after the geographic region that was devastated, this quake shook the ground much more violently than either the Haitian or the Christchurch earthquake had. The accelerations of the ground during the Tohoku quake are the highest ever recorded. Even so, there would probably have been relatively few casualties, except that the earthquake unleashed a second, and far more lethal, change of state. During the quake, one side of the earth's crust jumped 15–25 feet vertically in the blink of an eye (by geological standards), pushing a wall of water toward the coast—a tsunami. Reaching a height of slightly more than 132 feet (equivalent to a thirteen-story building) in one place when it reached land, and more than 50 feet in many others, the tsunami slammed into the coast, cresting over even the highest seawalls and devastating the coastal cities. The tsunami accounted for more than 90 percent of the 16,000 reported deaths. In some ways, this scenario of a tsunami triggered by an earthquake was a replay of events only seven years earlier when the magnitude 9.2 earthquake in the Indian Ocean generated a huge tsunami with waves 100 feet high. The waves stalked the whole Indian Ocean, killing more than a quarter of a million people in fourteen countries.

Half a world away, people in Iceland normally live in balance, though a precarious one, with fire and water—the fire coming from their volcanoes, and the water coming in the form of ice in their many glaciers. However, frequently the two come together to alter this normality. On March 20, 2010, melted rock ("magma") that had been rising, perhaps for years or even decades, into the plumbing system of Eyjafjalljajökull[3] (from here on, "Eyja") burst forth at the surface in a spectacular volcanic display (Figure 2.2).

Previous eruptions of Eyja had occurred over a two-year period between 1821 and 1823. Sheep and cattle died in the vicinity of the volcano, poisoned from gases now known to have been due to fluoride released during the eruption. But because the world was not globally interconnected then, the eruption was little noticed outside of Iceland.

FIGURE 2.2 Initial stage Eyjafjalljajökull eruption, March 2010. A line of vents has opened along a fracture, gas and ash are launched into low-level fountains, and lava flows are pouring off the flanks into the foreground. *Photo by Sigrún Hreinsdóttir, Institute of Earth Sciences (IES).*

The mountain is covered with an enormous cap of glaciers, but when the volcano burst to life in 2010, it broke out in an ice-free area on the flank. This eruption resembled many seen around the world in which gas propels pieces of red-hot magma and ash hundreds to thousands of feet into the air. This phase of the eruption was spectacular, yet mild enough that hikers and photographers climbed up to close vantage points to enjoy remarkable views of the activity.

Magma is melted rock that contains gases such as water, carbon dioxide, and sulfur dioxide. In its normal state in underground reservoirs, magma is like soda pop confined in a can: the gases are dissolved in melted rock in the same way that carbon dioxide is dissolved in flavored sugar water in soda. Energy is stored in the compressed gases. As the magma rose up in the plumbing system of Eyja, the pressure confining it decreased enough that the gases came out of solution, just as they come out of solution when a soda can is opened. When this happened, the state of the magma changed from a liquid containing dissolved gases to a liquid containing myriad individual gas bubbles, and then to a fragmented mixture of gas and hot ash that erupted into the air. As bubbles of gas burst out from the magma, they drove fingers of red-hot ash into the atmosphere—a change of state that resulted in a spectacular, but actually quite small, eruption. Although covered widely in the news (largely because the eruption was very close to a hiking trail and was accessible to journalists and tourists), this eruption caused almost no damage.

This style of eruption continued until April 12, when activity at Eyja ceased. But underground, new magma had found a path to the west where, only two days later, it erupted through new vents—into the ice-filled crater on the summit of Eyja. Magma poured onto ice, melting it and causing huge floods to run downward toward the southern coast. Eight hundred people had to be evacuated from the path of the floods. The meltwater also mixed with the hot magma, causing it to fragment into fine ash—a process much like what happens if you drop water into hot grease on a stove. Eventually, the magma itself erupted, propelled by hot gases into a towering plume. Such eruptions, first described by Pliny the Younger in the AD 79 eruption of Vesuvius, are called "Plinian eruptions," and their towering columns of ash and gas are called "Plinian columns" (Figure 2.3).

On the day of the renewed eruption, the jet stream happened to be flowing over Iceland. Ash rose to about 26,000 feet in the atmosphere—right into the jet stream. Simultaneously, a high-pressure system had parked itself over Europe, so the jet stream couldn't move. Eyja kept pumping ash from the eruption into the jet stream, which—like a gigantic conveyor belt—carried the ash into Scandinavia and Europe, right into the flight paths of major airlines.

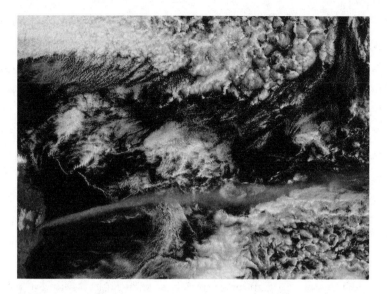

FIGURE 2.3 The brown, ash-laden plume from the second stage of the Eyjafjalljajökull eruption spreads from lower left to lower right out over the Atlantic, moving toward Europe. *Photo by NASA.*

The effect of even small amounts of volcanic ash on aircraft is enormous, as I saw firsthand while working at Mount St. Helens when it first came alive in the spring of 1980. We were helicoptering around the volcano in a snowstorm, unaware that the volcano was erupting a plume of ash that was mixed in with snowflakes. Upon landing, we discovered the effects of the ash: the small, but very sharp, particles had eroded all of the paint, and who knows how much of the metal, from the front edges of the helicopter blades! Ash can damage lights needed for landing on aircraft, can sandblast a plane's windscreen and force an instrument landing, and can damage the fuselage. It can clog the sensitive air intake ("pitot") tubes that are essential for measuring airspeed. Ash is like a ceramic and melts only at high temperatures, but jet engines are very, very hot. They are designed to cool by inhaling large amounts of air. If that air contains ash, it melts and then freezes onto sensitive parts, coating them with a thin ceramic film that changes their operating properties. In 1982, a British Airways plane flying at 36,000 feet went through an (invisible) ash cloud from an eruption in Indonesia, lost power, and plummeted from that altitude down to 12,000 feet—a terrifying vertical drop of 4.5 miles—before the engines could be restarted. In 1989, a KLM plane encountered similar problems after flying near Mount Redoubt in Alaska and suffered over $80 million in damage.

The personal and economic costs of the Eyja eruption in Iceland were enormous.[4] This relatively small eruption became the first large-scale twenty-first-century example of the vulnerability of our globally interconnected systems of transport and economics.[5] Airspace over Europe was shut down for six days, with more than 100,000 flights canceled over an eight-day period, affecting roughly 10 million passengers. Economic costs to the airline industry alone were estimated at $1.7 billion. European countries faced food shortages while African countries had to destroy food that could not be shipped. Asian economies were affected because supply chains for auto manufacturing were disrupted. All of this because disruption of the normal state of magma and glaciers released energy in a way that produced dynamic and violent processes.

Disasters like these are dynamic. They occur when energy stored in the earth in one form or another is suddenly unleashed in a way that harms humans. The release is episodic and often unpredictable. For a few brief moments of geological time, the normal processes of the earth seem to go crazy as their usually staid pace changes abruptly and briefly to a vigorous, forceful, and catastrophic tempo.

Disasters occur when something disturbs the status quo. That "something" differs dramatically from one event to another. In the case of earthquakes, stresses caused by motions in the earth's crust change the state of the earth around faults until finally they break. Energy is released during the quake, rocks around the fault quickly return to a new state of normality, and the cycle begins again. In the case of tsunamis, the normal state of the seafloor and ocean surface suddenly change when a fault breaks. The tsunami spreads out over the ocean, and in a few days the ocean returns to its state of normality. In the case of volcanic eruptions, gas long sequestered in magma is released, eruptions occur and then die or are extinguished, and the systems settle into their new states of normality.

These examples show that when something changes, disaster is a possibility—one that we need to keep in mind as we try to survive as individuals and as a species on this planet. Disasters are becoming ever more costly, largely because more and more people live in disaster-prone areas as our population soars. We cannot prevent Earth from occasionally unleashing its energies, but we can plan wisely, taking into account what will be required for remediation and recovery.

EXPLODING BICYCLE TIRES: CHANGES OF STATE

It's not very helpful to say, as I just did, that "something" changes and causes disasters. We can do better than that. More specifically, the broad view of the changes in the previous examples reveals two general categories:

+ First, the conditions of the *materials* change. Solid rock is ground up nearly to powder in earthquakes; solid earth, "terra firma," can turn to a liquid-like state. These are the types of changes that I refer to as "changes of state."

✦ Second, the conditions of *motion* change. Magma and
gas accelerate from rest to incredibly high velocities during volcanic eruptions. These changes are referred to by
fluid dynamicists as "changes of regime," a term unfortunately popularized in a different sense by Presidents
Bill Clinton and George W. Bush in the context of over-
throwing the regime of Saddam Hussein in Iraq.

The study of the dynamics of disasters is a study of how changes
of state and of regime take place, how the various forms of energy
in the earth change over time, and how the rates at which these
changes occur cause disasters.

Technically, changes of state and regime are described by a
set of parameters known as "state variables," and by equations
that relate these quantities. In mechanical systems, such as cars
moving along a freeway or planets orbiting the sun, typical state
variables are position and change of position (velocity). Thermodynamic systems are those in which temperatures change. In
thermodynamic systems such as expanding gases during bicycle
tire explosions or volcanic eruptions, typical state variables are
temperature, pressure, internal energy, enthalpy, entropy, and
volume. In these systems, the relations among the state variables are described by equations called "equations of state,"
and changes in the state variables are referred to as "changes
of state." A familiar example might be the equation of state of
a gas that relates the state variables of pressure, temperature,
and volume.

Motion is driven by forces and is described by positions, velocities, and accelerations. The equations that govern motion and
relate it to the forces acting on a system, and on its material properties, are called "conservation laws."

This business of equations of state and conservation laws may sound very complicated, but it's amazing how often they crop up in our daily existence. My dear husband isn't very mechanical, and he's not a scientist, but he inadvertently (and memorably) encountered changes of state and regime one day when he was pumping up a slightly deflated bicycle tire in our garage. Since it wasn't fully deflated, for most of this endeavor the tire was effectively a rigid reservoir with a constant volume (a bit over a gallon), simply acting as a container receiving gases at higher and higher pressures as he pumped and pumped away in the garage. Gases are compressible: when you apply pressure to them, their volume decreases (or, equivalently, their density increases). As he pressurized the tire, not only did the density increase, but so did the temperature. (You can feel this effect by placing your hand on a bicycle or car tire while inflating it.) The gas is obeying the equation of state that relates the three variables of volume, pressure, and temperature. When my husband was almost done inflating the tire, the gas reached a high enough pressure that some of the energy of his pumping was going into stretching the tire itself.

Unfortunately, my husband, who is very strong, didn't read the pressure limits printed on the tire, and he just kept on pumping and pumping until—BLAM! The tire burst. Even when his ears stopped ringing a few days later, he wasn't the slightest interested in my delight that he had discovered changes of state and conservation laws. The release of pressure allowed the gas to expand into the bigger volume of the garage, and this expansion drove the gas from rest in the reservoir to a very high velocity through the tear in the tire—a manifestation of the conservation laws of motion. (I actually got him somewhat more interested in the science of this event when I showed him pictures of a volcanic eruption that did the same thing; see Figures 2.2 and 2.3). The gas changed state as it expanded, and the rubber changed state from being a stressed whole tire to being pieces of a tire.

When the tire burst, energy stored in the compressed air in the tire changed to energy of motion as the gas escaped through the hole in the tire. The air cooled (a change of state) and accelerated (a change of regime). As the gas rushed out of the tire, it acted like a piston pushing on the air around it, and because it rushed out so fast, it generated a shock wave in our garage.

This example shows how changes in the state of materials cause changes in the state of motion. During these changes, energy stored in one form in a material that isn't moving—for example, as internal energy in pressurized gas—changes to energy in another form when the material takes off. There are many possibilities that we'll explore in this book. Disasters in general are caused when materials move—often, but not always, at high speeds. Earthquakes, landslides, tsunamis, volcanic eruptions, and atmospheric storms—big and small—are all examples of this change in energy from one state to another that sets materials in motion.

FOOTBALL, YOUR BANK ACCOUNT, AND CONSERVATION OF STUFF

Materials that move around in systems obey certain principles that are easily illustrated in human systems. Take, for example, people leaving a stadium after an event, exiting first to the parking lot and then onto streets and highways to get to their homes. The places they leave from or move toward are called "reservoirs." The substance that moves is called a "stock," and the movement is called a "flow." The stock (people) flows from one reservoir (the stadium) to another reservoir (the parking lot). Or the stock (cars) flows from the parking lot into the streets. A stock is a quantity that exists at a certain time, and a flow is a change in a stock over time.

Some reservoirs are bounded by well-defined physical walls, such as stadiums and parking lots. In this book we'll deal with fairly well-defined reservoirs when we discuss, for example, earthquakes, landslides, and volcanoes. But walls as tangible physical objects can't be drawn around many reservoirs. In weather systems, for example, air in our atmosphere flows from a high-pressure zone to a low-pressure zone. To a meteorologist, these high- and low-pressure systems are defined by specific criteria for pressures rather than by a physical wall, but the concept of a reservoir is equally valid. We'll deal with these more abstract reservoirs when we discuss weather, rivers, and oceans.

The stock, flow, and reservoir concepts are widely used in economics, accounting, business, and science. A stock is measured in units of "something," and a flow is measured in units of "something per time." For example, consider money that is being withdrawn from a bank account to buy groceries and pay household expenses. The stock is measured in dollars, and the flow is measured in dollars per month (or week, or year). The reservoir that holds the stock is the bank account, and the reservoir that receives the stock is a wallet or a cash register.

Unfortunately, stocks and flows each have various possible units of measurement, and this is often confusing. Stocks could be in pounds, tons, miles, or other units; flows could be in pounds per second, tons per minute, miles per hour, or other units. I will try to use intuitive units wherever possible, but always remember that we are basically just talking about "stuff" (the stock) and how it moves, or "stuff per unit of time" (the flow).

Three laws that can loosely be described as "conservation of stuff" govern how stuff moves. "Stuff" is mass or momentum or energy. These three laws must be supplemented by a description of the properties of the material that moves. Is it a gas? A liquid? A solid? Or a combination of these three?

Conservation of mass is a very simple concept: mass goes in somewhere, and mass comes out somewhere else. If the mass that goes in does not equal the mass that goes out, then mass has been stored in the space between where mass went in and where it came out. Mass can't just disappear. Think of your house: You bring stuff in from a shopping trip. You take stuff out to the garbage. If these two are not equal, then stuff ends up being stored in your house. You can generalize a bit to think of an assembly line into your house: Stuff comes in at a certain rate, which is the speed of stuff coming in—for example, pounds per week. Stuff goes out at another speed, also in pounds per week. If these two are not equal, then stuff ends up being stored in your house, again, as pounds per week.

Conservation of momentum is a bit more difficult. Momentum is mass in motion. Think of two football players—one big and fast, the other small and slow. Imagine that they both try to tackle a medium-sized quarterback. The little guy just bounces off the quarterback, whereas the big guy demolishes him. This example illustrates momentum—a combination of mass and velocity. The big guy simply has more momentum than the little guy, because he's bigger and because he's moving faster. Conservation of momentum is Newton's second law: a body responds to a force by changing its momentum. Material in motion changes its momentum in response to applied forces. In the football collision, the momentum of the tackler decreases as he hits the quarterback, and the momentum of the quarterback increases in response to the forces of the collision. The change in momentum of the whole system, consisting of the two players, is, however, zero (in the ideal case where no energy is lost in the collision).

Conservation of energy is an empirical observation that the total amount of energy in a system stays the same over time (is "conserved"), but can change forms.[6] The forms of energy of concern in this book are gravitational potential energy, kinetic energy, internal energy, and strain energy, often called elastic energy. "Potential energy" is energy that is available because something is high in the force field of gravity. If you are on top of a tall building, you have more potential energy than someone on the ground floor has. Should you jump off the building, the potential energy is transformed into "kinetic energy" as you accelerate in freefall. Kinetic energy is energy you acquire by moving; the faster you go, the more kinetic energy you have. The total amount of energy stays the same, but its form changes as you jump from the top of the building and fall toward the ground. If you are lucky when you jump, you might land on a trampoline, and then your kinetic energy will be transformed into "strain energy" deforming the mat of the trampoline. Strain means, simply, "a change of shape," and strain energy is energy obtained when the shape of an object changes because forces have been applied to it. If there is no trampoline, the kinetic energy is, unfortunately, dissipated in the fracturing of your bones, the sound of your impact, and heat. If you are carrying a soda bottle when you jump, and if this bottle breaks when you land, "internal energy" stored in the compressed gas in the soda is released and transformed into kinetic energy that sprays the soda everywhere. There are other forms of internal energy, but these are the primary ones that we will be most concerned with in this book.

To completely describe material motion using the conservation laws, we have to add a description of how materials respond to forces internally. For example, when you increase the pressure on a gas, it becomes denser and, under some conditions, warmer. The density of gas changes when pressure is applied or released, which means that the gas is "compressible." An equation of state describes mathematically how the state of a gas changes. The equation of state for liquid water flowing in channels is simpler than that for gas. For all practical purposes of this book, when pressure is applied to water the density doesn't change. Water is "incompressible": its material state doesn't change significantly in most problems (until it freezes or boils). Therefore, when we look at questions like the behavior of water, lava, or sometimes even air flowing in a channel, it is okay to ignore the equation-of-state issues, and it is sufficient to talk only about regime changes. We will explore these behaviors further in later chapters.

FLOWING RIVERS: CHANGES IN REGIME

Once materials are moving, the environment in which they move becomes important, as you can see by simply watching water flowing in a channel, such as a natural stream or river (Figure 2.4). The motion of materials depends not only on the forces acting on them, but on conditions when they start to move ("initial conditions") and conditions of the environment in which they move ("boundary conditions"). Take, for example, the motion of cars on a road that has two lanes in each direction and a stoplight. The stoplight has just turned from red to green. One car was at a complete stop, and the driver has to accelerate up to speed. Another car coming up from behind in the second lane never had to slow down and zooms on through the light. The cars have very different motions because their initial conditions when the light turned green were different.

Now imagine that there is a narrow bridge on the road and that the traffic must merge into one lane in each direction. The motion is different from that on the unconstricted two-lane road because the boundary conditions are different. Boundary conditions can force a change of state in a flowing fluid—a "regime

FIGURE 2.4 Regime changes in the pool-and-riffle configuration of a small stream. Water accelerates from
rest in a pool to higher velocities in a "riffle," and then
back to rest in another pool. This sequence of regime
repeats multiple times in some natural streams. *USDA
photo from http://www.ars.usda.gov/research/docs.
htm?docid=4098&pf=1&cg_id=0.*

change." We will see examples of regime changes in our discussions of wind, volcanic eruptions, and tsunamis.

Because they are inexorably intertwined, state changes and
regime changes pose a classic chicken-or-egg problem. On the
one hand, changes of state can cause changes in motion. On the
other hand, changes in motion can cause changes in state. In geological situations, it is often not possible to unravel which came
first. As we will see when we examine the dynamics of disasters,
when these changes occur on a geological scale, instead of the
scale of a garage, the consequences can be disastrous.

Chapter 3

WHEN TERRA ISN'T FIRMA

HAITI, CHRISTCHURCH, SHAANXI, AND . . . WASHINGTON, DC?

Earth's crust is amazingly strong—so strong that we take for granted our "terra firma." But in an earthquake, several changes of state cause terra to be not so firma at all, as residents of the Washington, DC, area found out in August 2011, when a relatively small earthquake of magnitude 5.8[1] rattled the corridors of power. Centered under a nearby small town in Virginia, and thus referred to as the "2011 Virginia earthquake," it caused no deaths and cracked only a few national monuments. But it brought home the reality of earthquake hazards to millions of people, as well as offering them a glimpse of the perils faced by hundreds of millions around the planet who live in geological settings where earthquakes a hundred or even a thousand times as powerful are part of the fabric of their lives.

On a late afternoon in January 2010, while residents of Port-au-Prince, the densely populated capital of Haiti, were going about their daily business, an earthquake of magnitude 7 struck only 15 miles away. Smoke and dust from falling buildings filled the sky. After twenty terrifying seconds, "there was nothing but rubble and dirt." Survivors spoke of "houses falling and falling, all of the fences were falling, people were falling."[2] Story after story from the survivors described destruction of buildings, broken water pipes, screams of victims alive but trapped in the rubble, and the recovery of bodies. As many as 300,000 people died,[3] and more than 1.5 million became instantly homeless, only to face months and even years of discomfort, disease, and death. In striking contrast, less than a year later an earthquake of similar magnitude (7.1) struck about 30 miles from Christchurch, a major city in New Zealand.[4] Only two people were injured (a third person died of a heart attack that may or may not have been caused by the earthquake).

Almost five centuries earlier, and half a world away, in a heavily populated area of Shaanxi Province, China, a magnitude 8 earthquake, nearly ten times as strong as the Haiti or Christchurch earthquake, killed over 800,000 people.[5] This number is equivalent to the entire population now living in San Francisco, California, or Austin, Texas.[6] Earthquakes killing thousands to tens of thousands of people occurred a number of times in the twentieth and early twenty-first centuries, and not only in China and Haiti, but also in the Middle and Far East. For us in the Western world, the loss of thousands in a single event is barely conceivable, and the loss of tens and hundreds of thousands is utterly inconceivable. Yet it happens every few years somewhere on the planet.

Some earthquakes cause the formation of huge tsunamis (the subject of Chapter 6) that kill a great number of people and destroy nearly everything in their path. The 1960 magnitude 9.5 Chile earthquake, the 1964 magnitude 9.2 Alaska earthquake, the 2004 magnitude 9.2 Sumatra earthquake, and the 2011 magnitude 9.0 Tohoku earthquake are recent examples. But tsunamis were not a factor in the cases of China and Haiti already cited.

What caused so many deaths in these earthquakes that didn't have tsunamis?

After weather-related events such as floods or blizzards, earthquakes are the deadliest of the natural disasters that occur fairly regularly. In addition to the human toll, costs of major earthquakes are measured in hundreds of millions to tens of billions of dollars. These costs are increasing as more and more people crowd into vulnerable urban areas on this densely populated planet. The global insurance industry is already reeling from the costs of earthquake-related disasters in the first years of the twenty-first century. Earthquake insurance in some regions, where available, is often prohibitively expensive. Because of the high cost, only 12 percent of home owners in earthquake-prone California have earthquake insurance.

Why are there so many fatalities in some earthquakes (for example, Haiti) but not in others of nearly identical magnitude (for example, Christchurch)? What happens deep in the earth during an earthquake, and how does this cause so much damage on the surface? Our field trip takes us briefly to Haiti, Sumatra, China, New Zealand, Japan, and Italy, with longer stops in New Madrid, Missouri, and Memphis, Tennessee. We will discover not one, but two changes of state that occur during many large earthquakes: crushing of solid rock, and a process known as "liquefaction." Our post–field trip reflection introduces the issue of communicating with the public about what we know and don't know.

EARTHQUAKES, VIOLINS, AND FURNITURE MOVING

One of the most important geological discoveries of the twentieth century was an explanation for the fact that Earth's crust (the top

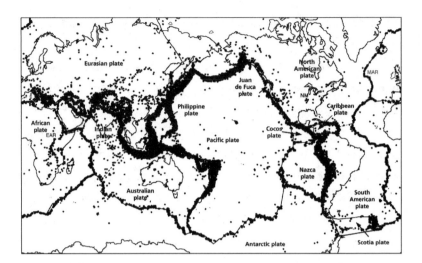

FIGURE 3.1 The major plates on the planet, as reflected in the locations of earthquakes, shown by dots. *From S. Stein,* Disaster Deferred *(New York: Columbia University Press, 2010).*

tens of miles in most places) looks like a jigsaw puzzle wrapped around a basketball (Figure 3.1). This explanation is the theory of seafloor spreading and plate tectonics. Eight major plates bump and grind around the planet (African, Antarctic, Eurasian, Indian, Australian, North American, Pacific, and South American), with an equal number of secondary plates, and another few dozen microplates filling the holes in between. In some places, such as the San Andreas Fault in California, the plates slide past each other. In other places, such as the west coast of Canada or the east coast of China, the plates plunge beneath, or ride over, each other. And in still other places, such as southwestern China where the Indian and Eurasian Plates meet, the plates collide and build huge mountain ranges like the Himalayas. The plate motions are driven by circulating currents in the asthenosphere, the weak zone of the earth in the upper mantle below the crust. The circulation of these currents drags the plates along like scum floating on a pot of boiling soup. Over the long 4.5-billion-year history of the Earth, plates have come together and stuck, old plates have been destroyed, and new plates have been born.

Throughout these intricate motions, the conservation laws of mass, momentum, and energy govern the dynamics. Internal energy stored in heavy elements in the earth is released by radioactive decay, providing thermal energy that drives the plates, kinetic energy in the motion of the plates, potential energy in the creation of mountain belts, and strain energy in the slow deformation of the plates. Strain builds up particularly near the plate margins, which are like elastic bands that store energy while being stretched, or like pencil erasers that store energy if being compressed.

Strain stored in the plates is released suddenly when a threshold of rock strength is exceeded and the crust of the earth breaks along a fault during an earthquake[7] (see Figure 3.2). Although most of the strain occurs at plate margins (see Figure 3.1), there are occasional earthquakes in the interior of plates, well away from modern plate boundaries—such as the Virginia quake. These earthquakes occur where old, and often hidden, plate boundaries are reactivated by the stresses generated by modern plate motions, or by other processes, such as the rebound of the crust from the unloading of ice at the end of the last ice age. When the ice was present, it pushed the crust down just like a mattress deforms under the weight of bodies, and when the ice melted, the crust rebounded. This process, which started about 14,000 years ago, probably contributed to both the Virginia and the New Madrid earthquakes.

Faults are not simple straight lines (Figure 3.2). On all scales, from global to microscopic, they zig and zag, and for this reason fault motion during an earthquake is like pieces of a jigsaw puzzle trying to move past each other. The irregular geometry of faults creates zones that respond to stress in different ways. Along

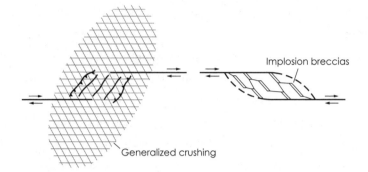

FIGURE 3.2 Schematic of a fault zone, in which one part of the earth's crust is moving to the right (*top*) relative to another part (*bottom*), along a fault. On the left, the relative motion creates a zone of crushed rock, a breccia. On the right, a gap opens and pieces of rock implode into this zone of tension, indicated by the open framework of the pattern in this region. *From R. H. Sibson, "Brecciation Processes in Fault Zones: Inferences from Earthquake Rupturing," Pure and Applied Geophysics 124, nos. 1–2 (1986): 182.*

some parts, the faults do simply slide past each other (Figure 3.2, middle). Along other parts they crunch into each other where rock on one side of the fault presents an obstacle in the direction that the other side is trying to move (Figure 3.2, left). And along yet other parts, where there are no obstacles to the movement, a gap can open (Figure 3.2, right).

The motion along a fault is spread out over a zone. The width of this zone depends on the geometry of the fault and how long the fault has been active. Almost all large earthquakes happen along preexisting faults, where the rocks in the sliding and the crunching parts of the zone have been repeatedly crushed and ground up to form a clay-rich product called "fault gouge" or "fault breccia." During an earthquake, old gouge gets reworked and new material adjacent to it also gets crushed.[8]

The motion of the plates is not a beautiful, smooth, and graceful ballet, but more like a raucous, grinding burlesque. Sometimes the plates stick together, and during these times stress builds up along the plate boundaries as circulation in the asthenosphere keeps trying to move the plates around. Other times the plates sprint—geologically speaking—past each other. This "stick-slip motion" is not unique to plate tectonics. It is in evidence any time you try to move a heavy piece of furniture: it takes a good heft to get the motion started, but considerably less force to keep it moving. Stick-slip motion also occurs when a violinist moves a bow across the strings of a violin. The motion of the bow drives the strings into cycles of stick-slip-stick-slip, hundreds per second depending on the string and note. In each cycle, the bow moves the string away from the equilibrium position during the stick phase, and then the string slips back toward equilibrium in the slip phase.[9]

Like the violin string, rocks on either side of a fault slip by each other in the slip phase when stresses exceed a critical level. The energy stored during the stick phase of the stick-slip cycle is released during the earthquake that accompanies fault slip. This process repeats over and over as different parts of a fault break. The recurrence of major earthquakes in cycles of a hundred to a thousand years reflects the fact that energy is built up and stored for centuries before being released in just a few seconds or minutes during an earthquake.[10]

SHAKE, RATTLE, AND ROLL

During an earthquake, changes of state are not limited to the fault zone itself. Energy is radiated away from the fault by waves, some of which we feel as vibrations and shaking of the ground. These waves disperse the concentrated energy released at the fault to distant regions of the planet. They compress and shear materials along their path as they travel the Earth (body waves) and over its surface (surface waves). Because body waves propagate through the whole Earth, they provide seismologists a valuable tool for unraveling the internal structure of the Earth. Surface waves, in contrast, are responsible for most of the destruction. They travel more slowly than body waves and arrive at any particular site well after the body waves have passed. Thus, thanks to the speedy body waves, people sometimes get enough notice of an imminent earthquake to take cover before the arrival of the slower, more destructive surface waves.

I was in Los Angeles when the 1971 San Fernando Valley earthquake struck early one morning. Our four-year-old son was asleep in another room. At magnitude 6.6, this earthquake was of only moderate size, but it killed sixty-five people and caused more than a half billion dollars of damage. It was sufficiently notice-able to inspire one of Hollywood's earliest disaster blockbusters, the 1974 film *Earthquake*. The shaking lasted about ten seconds, long enough for us to wake up and leap out of bed to get to our kid, only to find that we were completely disoriented and couldn't do anything but try to keep our balance! The quake was over before we had moved even a few feet away from the bed—and our kid slept through it!

A larger quake unfolds like the San Fernando Valley quake, but on a longer timescale and with more vigorous shaking. The intensity builds up over seconds to tens of seconds. People stagger and reel, struggling to stay connected to the ground without being flattened onto it. Thrown to the ground, most are unable to get back to their feet. A deafening roar builds up. In cities, goods fly off of shelves, pieces of buildings start to break off, and whole buildings start to collapse.

Buildings in many places might withstand a single sharp jolt, but only buildings specially engineered to earthquake safety standards can withstand prolonged shaking, which sometimes lasts for many minutes. During the 2010 Haiti earthquake, a rupture started 5 miles underground, displacing the Caribbean and North American Plates by at least 13 feet. The rupture was over in about ten seconds, but the shaking lasted nearly a minute.[11] Although well-engineered buildings might have withstood this shaking, Haiti is one of the poorest and least developed nations in the world and is plagued by bad construction practices. Contractors used the wrong kind of steel and concrete in buildings, they mixed cement with dirty or salty sand, and shoddy construction practices resulted in weak buildings.[12] In contrast, the Christchurch earthquake lasted about forty seconds, but damage was minimal because strong building codes existed and were enforced.

The rupture event during the Tohoku earthquake lasted about two and a half minutes.[13] During this time the faults separating the Eurasian and Pacific Plates slipped as much as 130 feet across an area only slightly less than the combined areas of Vermont and New Hampshire.[14] It was a complicated sequence of events in which probably three separate ruptures broke different parts of the plate interface at different times. Each subsequent event started at the boundary with the previous rupture and propagated into regions that had not yet slipped.[15] Had the whole area ruptured in one single event, the magnitude 9.0 earthquake would probably have been an even more devastating magnitude 9.4.

The record for the longest duration of an earthquake belongs to the 2004 Sumatra quake (Figure 3.3). During this earthquake, an area[16] comparable to that of Oregon or Wyoming moved as much as 36 feet along the fault that separates the Indian and Eurasian Plates.[17] The event unfolded over ten minutes as the rupture propagated northward along the fault.

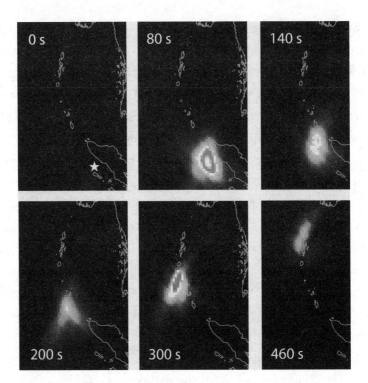

FIGURE 3.3 The progression of the rupture from the Sumatra earthquake. The time after the beginning of the event is
shown at the top of each frame, in seconds. The light patches
are related to the energy released in the seismic waves. The
north tip of Sumatra protrudes from the lower right, and the
Andaman and Nicobar Islands run up the middle of each
frame. *Modified by M. Ishii from M. Ishii et al., "Extent, Duration
and Speed of the 2004 Sumatra-Andaman Earthquake Imaged
by the Hi-Net Array," Nature 435 (2005): 933–36. Used with permission from Macmillan Publishers Ltd: Nature, copyright 2005.*

Shaking caused the deaths in the deadliest earthquake yet
recorded, the 1556 Shaanxi earthquake in China, where tens
of millions of people have lived for centuries in elaborate caves
("yaodongs") carved into enormous deposits of windblown dust
(loess) that cover much of central China.[18] Yaodongs have only
the strength of weak clay and couldn't withstand the vibrations

from the magnitude 8 quake. Many collapsed, killing more than 800,000 people, an estimated 60 percent of the population in some areas. Since the loess deposits cover an area the size of Texas (6.6 percent of the area of China) and are in an area of large earthquake hazards, research on the way loess responds to shaking has a very high research priority among Chinese engineers.

SHAKE, RATTLE, AND JIGGLE

The first of the two changes of state that I look at in this chapter can take place up to hundreds of miles from the epicenter, where the shaking, though not severe, is strong enough to cause a change of state when the ground consists of wet granular materials, such as wet soils, sandy deposits, or river muds and gravel. In places, water suddenly starts to appear on the surface, and then geysers jet up tens of feet, often carrying mud, soil, and stones or pieces of wood.[19] Where the surface deposits are dry or only partially wet, air roars and whistles as it is exhaled from underground voids. In wooded areas, trees sway, crack, and fall into tangled masses. Other times the dewatering process is more gentle, but either process often leaves large muddy deposits on the surface and, in some places, craterlike features known as "sand blows" or "sand volcanoes," layers of sand that slope away from a central vent marking the channel where water emerged to the surface. Sand volcanoes are typically a few feet in diameter, but they can cluster along fractures and cracks in the ground, forming fairly big fields of sand and mud. When all is done, the landscape over quite a large area can be covered with water and turn into mud (Figure 3.4).[20]

Historically, this process, now known as "liquefaction," has contributed to the deaths of hundreds of thousands of people, but widespread awareness of the process didn't begin until roughly

FIGURE 3.4 Soil liquefaction in Pajaro, California, after the Loma Prieta earthquake, October 1989. The sand-volcano vent in the foreground is about 4 feet in diameter. *USGS photo by J. C. Tinsley.*

200 years ago, and it has continued to increase up to the present time. In 1811–12, when a young Davy Crockett was a scout with the Tennessee militia and an aging Daniel Boone was hunting and trapping in the hills of Missouri, a series of earthquakes, now known as the New Madrid sequence, rattled Missouri, Tennessee, Kentucky, and the southern parts of Illinois and Indiana. Four earthquakes, and many aftershocks, occurred along a fault now identified as the Reelfoot Fault. The earthquakes occurred at a time when the Mississippi River was briefly the western frontier. Had they occurred a century earlier, there would have been no (European) eyewitness records;[21] and had they occurred one or two centuries later, they would have caused much more damage.[22]

Soils are normally strong enough to support even heavy buildings because the individual particles in the soil are in contact with each other (Figure 3.5*a*). The weight of the particles on the areas that are in contact holds the particles in place and gives soil its strength. Generally, water in the pores of soil is under rather low pressure and doesn't affect the contact areas of the grains. However, if the pressure on the water increases, as it can during an earthquake (or during blasting operations with dynamite), it can exert pressure on the soil particles, forcing them farther apart and reducing, or eliminating, the contact areas (Figure 3.5*b*). (If you have ever played in quicksand in a sandbox or at the beach by jiggling your hands or feet, you have produced this same effect by disturbing the water.) The high water pressure in the pores reduces contact areas of the grains and, thus, the strength of the soil. In extreme cases, the water pressure can become so high that the particles are just floating in the water and the soil has almost no strength at all; it acts like a liquid and can't support any weight. The transformation of a soil from its normal strong state to a nearly liquid state can be nearly instantaneous.

Liquefaction can distort the ground, often causing spreading that results in settling and fissuring. Downward slopes and hillsides in the area often show pronounced slumping. Liquefaction can occur in small local areas, producing an isolated feature, or over large areas, producing extensive modification of the surface. It results in settling, with entire buildings moving sometimes uniformly but other times differentially, in which case the buildings end up severely tilted. It also causes shifts between foundations and the structures they support. In the central business district of Christchurch, liquefaction during the February 2011 earthquake caused the collapse of two multistory buildings and the collapse or partial collapse of many unreinforced masonry structures, including the famous Christchurch Cathedral. The central part of the city was, for all practical purposes, devastated.

Even though the New Madrid events provided irrefutable evi-

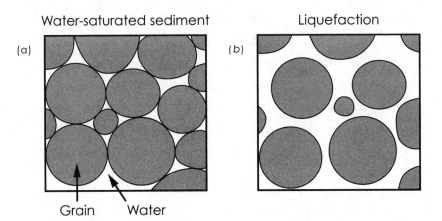

FIGURE 3.5 The configuration of sand grains in water-soaked ground in their normal state (a) and during liquefaction produced by shaking (b).

dence of the role of liquefaction in destruction during earthquakes, the frontier nature of the central US and lack of global communication about events in other countries hindered the study of this process. Things began to change in the mid-twentieth century. Studies of the 1920 Haiyuan earthquake that killed more than 230,000 people in China showed that liquefaction of water-saturated layers in the loess had triggered large, high-velocity landslides that caused the enormous amount of devastation and high number of fatalities.[23] The magnitude 9.2 earthquake in Alaska in 1964, the magnitude 7.5 Niigata (Japan) earthquake the same year, and the 1983 Japan Sea earthquake brought liquefaction to widespread attention of the public, scientists, and engineers. In these earthquakes, liquefaction played a major role by undermining structures, and triggering and lubricating landslides. In general, liquefaction can cause masonry and wooden houses to collapse; destroy earth dams; cause road and railway accidents; float buried structures such as sewage treatment tanks;

destroy irrigation and drainage channels; and sink, tilt, or uplift wooden poles, such as electric or telephone poles, thus also causing dislocations to communication infrastructure at a critical time of emergency.[24] In the 2011 Tohoku earthquake, water, sewer, and gas pipelines were destroyed, crippling critical infrastructures over hundreds of square miles. In major earthquakes (magnitude greater than 8.0), liquefaction has been observed up to 250 miles from the epicenter.[25] The US military has concluded that in some cases liquefaction accounts for more damage to their installations than does the actual seismic shaking of structures during earthquakes.[26]

Global communication had improved so much during the last decades of the twentieth century that satellite photos of liquefaction in Haiti and YouTube videos of liquefaction during the Tohoku earthquake were posted and distributed widely within days of those earthquakes, bringing the phenomenon vividly to life for a wide audience.

Many countries, including the US, New Zealand, and Japan, have instituted and enforced strict building codes in areas identified as hazardous because of potential liquefaction. Unfortunately, enactment and enforcement of strict building codes in earthquake country is not universal. Liquefaction, combined with poor construction practices, caused much of the devastation of the harbor in Port-au-Prince in the 2010 Haiti earthquake. The difference in death tolls between Haiti and Christchurch is directly attributable to different codes and practices, as well as the degree of enforcement of the codes.

Poor decision-making and construction practices are not unique to poverty-stricken countries. They occur in the US as well. Memphis, Tennessee, a city that could see strong ground shaking if there is another earthquake like the major one of the 1811–12 New Madrid quakes on the Reelfoot Fault or other

FIGURE 3.6 The Great American Pyramid of Memphis, Tennessee, a 1:60 scale replica of the Great Pyramid of Egypt. *Photo by Thomas R. Machnitzki.*

faults in the region, boasts of "The Great American Pyramid" (Figure 3.6). Standing 321 feet tall and capable of holding 21,000 people, the structure was completed in 1991, but unfortunately, only in 1992 did the city begin thinking about earthquake construction standards. The pyramid sits on artificial fill, mostly silt and sand laid down along the banks of the Mississippi River.[27] The predicted shaking in the event of another big New Madrid earthquake would cause moderate to heavy damage.

SHAKE, BAKE, ZAP, AND GLOW

The second of the two changes of state that I examine in this chapter occurs right in the fault zone itself during the earthquake. Much less is known about this change of state than is known about liquefaction. In the fault zone, a major change of state transforms the rocks in a way that facilitates their sliding past each other. Hints about this process are found in rocks and, of all places, in the sky. An old Japanese haiku speaks to an optical phenomenon that accompanies earthquakes:[28]

> *The earth speaks softly*
> *To the mountain*
> *Which trembles*
> *And lights the sky.*

One of the "darkest areas of seismology"[29] is a phenomenon known as "earthquake lights," or sometimes "earthquake lightning." This phenomenon has a variety of manifestations. In many cases the sky lights up in a way that resembles sheet lightning during hot summer months. Sometimes there is a flash of lightning that seems to last much longer than normal storm lightning, but no thunder. The sight has been likened to the aurora borealis with streamers diverging from a point on the horizon. The lights may appear like beams of Hollywood searchlights piercing the night sky, tongues of fire, or flames emerging from the ground. Some witnesses describe globular incandescent masses, like "fireballs" (Figure 3.7). Broadly, the phenomena are described as "electromagnetic" because it is believed that electrical and magnetic forces are at work.

Earthquake lights have been described anecdotally for centuries, and they have even been captured on film and video,[30] but there is still no broad consensus about their cause or their usefulness for earthquake prediction. Different theories have suggested that the lights are produced by heating caused by friction, by emission of radon gas released during fracturing, by generation

FIGURE 3.7 Eyewitness sketch superposed on a photo showing a luminous phenomenon (earthquake lights) observed at the time of the 1911 Ebingen earthquake. The lights started with a bright flash from the ground that turned into a large luminous sphere at a certain height. It lasted a few seconds before dividing itself into lightning-like sparks. *A. von Schmidt and K. Mack, "Das süddeutsche Erdbeben vom 16 November, 1911," Abschnitt VII: Lichterscheinungen, Württembergische Jahrbücher fur Statistik and Landeskunde, Jahrg., part 1 (1912): 131–39. Kindly provided with background information by John Derr.*

of light or electric charges from strain in the rocks, or by electric charging of fluids moving through cracks—but no theory has garnered a significant consensus. Rocks recovered from ancient fault zones now exposed at the surface may provide a clue.

A spectacular black rock, chock-full of fragments of the surrounding rocks that are often pink or white, is found in some fault zones (Figure 3.8). Since this rock resembles the volcanic rock known as tachylite, it was named "pseudotachylite" (sometimes spelled as pseudotachylyte, and pronounced "soo-doh-TAK-uhlyte").[31] Pseudotachylites have been dubbed "flight recorders" because, like flight recorders recovered after aircraft accidents, these rocks tell geologists something about the conditions of their formation during a catastrophe. They've also been called "fossil earthquakes"[32] because they are found in old fault zones that have been exposed by erosion and they preserve a record of processes that occur in modern faults at depths too great for us to sample the rocks.

Ironically, given the "fossil earthquake" nickname, pseudotachylite was first discovered not on an earthquake fault, but in the Vredefort structure, South Africa, produced by an ancient meteorite impact.[33] During impacts, the kinetic energy from the speeding projectile is deposited at the impact site, where it is converted into kinetic energy of excavated ejecta, thermal energy in heated and melted rock, and seismic energy in ground shaking. Even though the geologist who discovered and named pseudotachylites did not know that Vredefort was formed by such a violent process, he concluded that the bizarre rock was formed by a "sudden gigantic impulse or series of impulses."[34] The only known occurrences of pseudotachylite are in structures associated with violent events: impact structures, large and catastrophic landslides, and earthquake fault zones.[35]

Pseudotachylites show evidence of both crushing, which occurs at fairly low temperatures, and melting, which occurs at much higher temperatures.[36] This dual evidence tells us quite a bit about what happens during an earthquake. Rock in the fault zone is first crushed into fairly large chunks. As the earthquake continues, the chunks are ground up, producing ever-smaller rock fragments, mineral grains, and crushed mineral grains. This crushing process generates the pieces of rock that remain in the pseudotachylites.

During fault motion and the crushing process, heat is gener-

(a)

(b)

FIGURE 3.8 (a) A sample of pseudotachylite from
a shearing and melting during a meteorite impact.
(b) Similar material is found in fault zones exposed by
erosion. *Part (a): photo by the Smithsonian Institution;
part (b): photo by Steve Marshak.*

ated just as it is when you rub your hands together on a cold day. This friction causes the rocks around the fault to become warmer and warmer. At the same time, heat flows away from the zone of crushing toward colder rock on the margins of the fault. If the heat generated exceeds the amount that can be conducted away from the fault, crushed material in the fault zone begins to melt. At first there is only a small amount of melt. But the melted material can become so hot that it melts and digests some of the smaller rock fragments. If you want a good analogy, put a bunch of crushed ice in a pot on the stove and turn on the heat. Small ice particles will melt first, and the liquid water produced by this process will melt and digest bigger and bigger pieces of ice.

Like water, melted rock has no strength. The generation of rock melt by motion of the fault makes it easier for the fault to slip more. The friction of that motion, in turn, generates more rock melt, so it's a runaway process. The process can be stopped in two ways: One is that the fault meets an obstacle that is so stiff that it can't be broken. The other is that so much crushed rock is generated that the melted rock cannot digest it all, so the fault lubrication stops. Once the motion along the fault ceases, the rock melt cools, producing a glass that contains the surviving rock fragments. (In Figure 3.8, the dark material is the melted rock, and the light-colored bits are the pieces of the original light-colored rock that remain. However, in other rocks the melt and the rock fragments may be different colors.) These rocks are evidence of massive, nearly instantaneous changes of state deep in the earth.

How can pink and white rock turn into black rock in just a few seconds or minutes? There are two culprits. The first is that even light-colored materials can appear dark when they are crushed into very small particles, because the reflectivity of light is reduced. The second culprit is actually more interesting. In many rocks, iron is present, tied up in the molecular structure of minerals. This iron is set free during the melting. Once freed from the original minerals, the iron reacts with oxygen, forming minute nanograins of black magnetite (which has the chemical formula Fe_3O_4, where Fe stands for iron, and O for oxygen) or other iron minerals only a few ten-thousandths of an inch in diameter. These nanograins give the rock melt its black color.[37]

But now the mystery deepens! The tiny magnetite crystals preserve a record of the magnetic field that existed when they were formed. Amazingly, the magnetic fields stored in these crystals are about a thousand times stronger than the Earth's normal magnetic fields. Magnetism can be created by strong electric currents, so the existence of strong magnetization in the rocks implies that strong local electric currents existed at the time of the melting. The magnetic fields created by these currents are imprinted in the rocks as their temperatures cool down below 1,000°F. The imprinted field is then preserved as the rocks cool down. The magnetic fields probably last only a very short time, perhaps several minutes at most.[38] The most likely causes of magnetic fields along the fault plane are strong electric currents that last only as long as the fault moves—a theory suggesting that a fault zone is like a giant car battery.

I think of myself as a very self-sufficient, tough geologist who can deal with flat tires and all sorts of car maladies in the field, but dealing with putting clamps on a dead car battery to jump-start a car always terrifies me. Imagine, now, your car battery scaled up to the size of a fault zone. In one model of pseudotachylite formation, the Earth is like such a battery with one unconnected terminal.[39] In this model, the crust of the Earth stores up energy that will be discharged through the equivalent of clamps on a car battery. When a rupture occurs along a fault and friction melt forms, the melted rock provides a conductive path that closes the circuit; the generation of melted rock is the equivalent of putting the second clamp onto the car battery. The electrical conductivity of a pseudotachylite melt is about 10,000 times larger than that of the host rock, so it acts like an ephemeral lightning rod. The current persists as long as the melt exists to keep electrical conductivity high. That electric current magnetizes the nano-magnetite crystals that are found in the pseudotachylite.

Because pseudotachylites contain so much melted rock, some scientists have suggested that a lot of heat is generated during earthquakes.[40] If this is the case, it follows that temperatures in active fault zones should be very high. One estimate is that the amount of energy released during the 1994 magnitude 8.2 Bolivia earthquake could have generated temperatures of 9,000°F.[41] But this remains a controversial subject because extensive studies on the San Andreas Fault in California fail to reveal such heat. Among other things, this conundrum has led to lively and prolonged debate about how the San Andreas works.

REFLECTIONS: RARE EVENTS, HIGH STAKES, AND RISK COMMUNICATION

Our knowledge of earthquake dynamics increased enormously during the twentieth century through seismic and other methods of monitoring fault zones, laboratory experiments on rock strength and behavior, and theoretical modeling of the dynamics of earthquakes. The US Geological Survey (USGS) now issues maps of earthquake intensity hazards using predictions of ground acceleration around faults.[42] However, earthquake prediction is, and probably will remain, imprecise, and communication of this risk is difficult—much more difficult than forecasting something like the movement of a tornado or a tropical storm.

The issues of known knowns and known unknowns are always with geologists as we try to understand, predict, and communicate the science of disasters. Two tools that we use are statistics and probability: How often does an event of a certain size occur? Once a year? Once a decade, century, millennium? Once every 500,000 years? Often we are dealing with events that have a low probability but high consequences. We will encounter them in every chapter. Humans seem hardwired to deny the existence of these events, and as a result, we are almost always underprepared for the rare, big event. Because there are so few of the truly rare, but plausible, events, the methods of statistics, which rely on having a large data set to analyze, are of little help. Other methods of studying these events must be applied—such as detailed modeling of the governing dynamical processes. Even though such events have a small probability of occurrence at the timescales of our lives, or even the timescales of our civilizations (a few tens of thousands of years), we cannot afford to ignore the dynamics—that is, the science— because just one such event may cause us great harm.

Geologists are human beings with roles not only in science, but as citizens as well. In our role as scientists, we evaluate the evidence and the uncertainties in it. In our role as citizens or advisers to the government, we provide our best science to officials who are charged with conveying the information to leaders and to the public, and with formulating policy and action.

Effective communication about these rare events to leaders and to the general public, whether by geologists themselves or by their representatives, is extremely difficult because humans have very little intuitive feel for low-probability events, and very little means of converting numbers into action. If you lived on a coastline and learned there was a probability that a tsunami would sweep away your house roughly every thousand years, would you sell it immediately and move away? If something has a probability of 0.00015, or 0.015 percent, what does that mean? That's your probability of dying in a car crash in any one year, but does that number affect anyone's personal decision to use or not use a car? The probability that an earthquake of magnitude 7.5 or higher will strike somewhere in California in a given year is much, much greater—about 2 percent. That number significantly affects the insurance industry, but it probably doesn't influence many people's decision about whether to live in California.

In the US, two models for risk communication can be found, one in the weather community and one in earthquake-prone California. A major lesson learned over many decades of experience with earthquakes of many sizes is that you cannot convey risk only when there is a crisis. Rather, information and education about the issues need to be given to people frequently and continuously. In this way, people get used to how Earth's dynamics change, as well as to controversies, false alarms, and changing conditions. As a result, citizens can develop mitigation plans on both personal and community levels. Individually, for example, people can build or reinforce (retrofit) housing to safer levels, decide whether to buy insurance, and stock up on food staples and water. Or not. Communities can adopt and enforce building codes or usage zoning, and define areas at risk—sandy areas vulnerable to liquefaction, low areas along the coast prone to tsunami damage after an earthquake, and hilly areas exposed to landslide hazards (see the next chapter.) Such areas can be dedicated to appropriate uses. Or not. Successful living in hazardous areas depends on scientists, engineers, and policy makers working together to formulate plans, and on public, political, and financial will to implement and to live by those plans.

Chapter 4

THE FLYING CARPET OF ELM

A FLOATING FARM AND
A FLYING CARPET

In 1978 a hapless farmer in a small town in Norway discovered what happens when liquefaction occurs on a hillslope.[1] His farm was set on a very gentle slope along the coast of Lake Botnen. Eleven thousand years ago, an ancient sea that developed at the end of the last ice age covered this area. As the glaciers advanced about 14,000 years ago, they ground rocks in their path to very fine powdery silt and clay. Then, as the climate warmed and the glaciers melted, water carried the clay and silt particles into the fjords at the sea. There the particles fell out of suspension in the water to form muds on the bottom of the sea. Chemical reactions between the muds and the salt in the seawater strengthened the muds—a process rather like the setting of freshly poured, watery cement. Later, after the weight of the glaciers had been removed, Scandinavia rebounded upward, elevating these cemented muds as much as 600 feet above sea level.

Then, over the next thousands of years, more reactions between

the excavated muds and rainwater transformed the top 20 feet of
these clays into a stiff layer suitable for building and farming.
At the same time, groundwater circulation deeper in the deposits
leached the salt out of any seawater that had been trapped in the
pores of the clays. And therein lies the rub, and the problem for
the hapless farmer: When the salt concentration of water in the
pores of clay-rich sediments drops below a critical threshold (to
about 0.1 ounce, only the weight of a spice package, per gallon of
water), the clay takes on an amazing character. At one moment
it can be strong, almost like a solid, but the next moment, under
certain conditions of stress, the state changes nearly instantaneously into a liquid. This liquefied material, "quick clay," resembles its more famous relative quicksand, except that the particles
are smaller. In Norway the areas underlain by quick clay consti-
tute some of the best farming land, so they are heavily populated.

In order to construct a new wing on his barn, the farmer had
removed 900 cubic yards of soil (equivalent to a cube measuring
about 30 feet on a side) and piled it aside at the shore of Lake
Botnen. That's when disaster struck.[2] The slight redistribution
of weight by the farmer from the barn area to the shore of the
lake triggered a slide of over 7 million cubic yards (equivalent to
a cube more than 600 feet on a side). The relatively small volume
removed by stockpiling paled in comparison to the amount of
material in what is now known as the "Rissa slide," by a ratio of
about 8,000 to 1! During several pulses, material flowed inexorably toward the lake over a period of five minutes. A witness said
that one pulse "came towards me like a sea wave."[3] Immediately
after this pulse stopped, a new and larger flake broke off, and a
video of the event shows the farm floating "in a matronly state"
down the quick-clay river at a speed of over 20 miles per hour.[4]
Forty people were caught in the slide, one died, and seven farms
and five homes were destroyed or abandoned.

Many regions formerly occupied by ice-age glaciers are vulnerable to quick-clay slides, particularly in Norway, Russia, Finland, Sweden, the US (Alaska), and Canada. In 1893, a midnight quick-clay landslide in Verdal, Norway, killed 116 people, nearly half of the community of 250. Even though quick-clay slides move fairly slowly in comparison to some other landslides, they are often deadly because they can occur without warning. A quick-clay slide in Saint-Jude, Quebec, in 2010 happened so suddenly that the members of one family died where they had been sitting—watching an ice hockey game on television.[5]

During the Rissa slide, fine-grained materials flowed down a gentle slope, but at the other extreme, some landslides consist mainly of huge rocks that roar down steep slopes. The small village of Elm sits below a mountain in an area of Switzerland where slate, a rock that fractures nicely to expose flat surfaces that make great blackboards, was quarried with explosives for decades. The introduction of obligatory public education in the mid-nineteenth century greatly increased the quarrying activities. In the late 1800s, cracks developed in the mountain, but the quarrying was too lucrative to give up in spite of these ominous signs, and so it continued—even after an exceptionally large crack opened in the rocks above the village in 1876. For five years not much happened, but then on September 11, 1881, after two months of heavy rain, a giant slab of the mountain disintegrated (Figure 4.1). For about twenty minutes, rock rained down on the village of Elm, culminating in 10 million cubic yards of rock roaring into the valley, across its floor, and 300 yards up the other side. One hundred fifteen people were killed.

When I was a child, I once tried to escape from the house unnoticed by my parents by climbing out my bedroom window, on the second floor, onto a roof over a porch so that I could jump to the ground, a good 10 feet below. Like most kids that young, I had

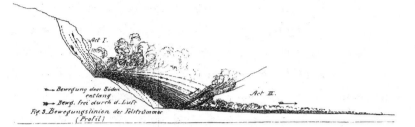

FIGURE 4.1 Illustration of the eyewitness sighting when the rock mass jumped away from the ledge in the 1881 landslide in Elm, Switzerland. The eyewitness saw the alder forest through the gap below the ledge, under the debris flying overhead. *From A. Heim, "Der Bergsturz von Elm,"* Deutsche Geologische Gesellschaft Zeitschrift 34 (1882): 74–115.

to learn the hard way that my knees would stop before my chin, and I nearly broke my neck when the two collided. An eyewitness described a similar phenomenon when material falling from the mountain above Elm hit a ledge (my knees), except that the later-arriving material (my chin) didn't land flat on the ground like I did, but went airborne like a flying carpet:[6]

> *Then I saw the rockmass jump away from the ledge. The lower part of the block was squeezed by the pressure of the rapidly falling upper part, became disintegrated and burst forth into the air . . . The debris mass shot with unbelievable speed northward toward the hamlet of Untertal and over and above the creek,* for I could see the alder forest by the creek under the stream of shooting debris. *(emphasis added)*

The floating farm and flying carpet illustrate several different changes of state in both materials and motion. To understand the incredible diversity of landslides, our field trip will take us around the world to Norway, Switzerland, China, Italy, and Alaska, with longer, focused stops in California and Wyoming. We'll even pay a brief visit to Mars! Finally, in our post–field trip meeting I reflect further on knowns and unknowns and the difficulties of communicating them to the public.

DIVERSITY: LANDSLIDES HAVE IT

Landslides are like people in at least one way: every one is different. They occur in settings ranging from jungles to deserts. The materials involved range from mud to rock to ice, including mixtures of all three. Some slides are wet; others are dry. Some roar down steep mountainsides; others creep along at barely perceptible rates. Some are so rigid that they are truly "slides"; others, so fluid that they are best described as "flows."

But the motion of landslides, small or large, is always governed by the conservation laws. Even when a hillside is static and stable, gravity is tugging everything downward. Usually, hillsides are strong enough to resist the pull, but all materials have zones of weakness, and when one of these zones fails, disaster looms. Initially, landslides flow downhill in response to gravity, although they can, and do, go uphill during parts of their travel if their momentum is sufficient to carry them upward. If a landslide were to fall straight down a vertical cliff, like me jumping off the porch roof, then it would accelerate in response to the full force of gravity. The potential energy that the mass of earth had high on the slope would be converted to the kinetic energy of motion as it fell. However, most landslides are not tumbling down vertical cliffs but are sliding down slopes, sometimes very gentle slopes that are more like very gentle slides on a playground. The force of gravity driving them is reduced by an amount proportional to the slope.

Conservation of momentum in the form of Newton's second law (force = mass × acceleration), as encountered in high school physics, determines the details of slide motion.[7] This deceptively simple equation, however, is actually very, very complicated because many forces act on a landslide. Specifying them all and adding them together is usually difficult. The major force at work is gravity, but it has a highly effective opponent: friction. At the base of the slide where it moves across the ground, friction between the slide and the ground opposes the motion just as it slows a book when you try to slide it across a table. As well, within the slide materials shear and deform, creating friction internally. The balance between gravity driving the slide downhill and friction resisting this motion determines the speed of the landslide and, eventually, how far it will travel.

The wide variety of landslide settings and material properties means that the physical laws of conservation of mass, momentum, and energy take on wildly different forms from one landslide to another.[8] All landslides start with a reserve of potential energy. In some, such as the flying carpet, conversion of potential to kinetic energy dominates. In others, such as slow-creeping landslides, dissipation of energy in friction dominates.

Reliable statistics on landslide fatalities are difficult to obtain, but by conservative estimates, nearly 90,000 people died in landslides between 2002 and 2012, an average of 9,000 people per year.[9] During these years, two catastrophic slides caused by earthquakes accounted for a large fraction of the fatalities. The 2005 magnitude 7.6 earthquake in Kashmir killed at least 30,000 people; the 2008 magnitude 7.9 Szechuan earthquake killed at least 25,000. Historically, earthquake-induced landslides have killed hundreds of thousands at a time in China and the Far East, and destroyed vast areas of agricultural land. Even when earthquake-induced landslides are removed from the statistics, an average of

4,000 people per year are killed in landslides, with variations that probably depend on El Niño and La Niña conditions not yet well documented.

Landslides descending from hillsides into river valleys often block the river channels, forming a dam behind which the water from the river ponds to form a "quake lake." In 1786 a magnitude 7.8 earthquake in southwestern China triggered a landslide that blocked the Dadu River.[10] About 100,000 people drowned in the flood. Although the US has far fewer fatalities from landslides than does China, the annual cost of landslides here still amounts to billions of dollars.[11]

Generalities are difficult to come by, but we can learn much about landslide dynamics by looking at several examples: a small, slow, and creepy one; a couple of small, energetic ones; and then the mother of all landslides. In so doing, we'll discover why there are enormous challenges in predicting when and where land-slides will occur, how far they will travel, and how they will travel. These are important issues because human use of the land influences landslide dynamics, so if we are to minimize our own effects, we need to know what triggers slides and how they move.

A DANGEROUS COMBINATION: GEOLOGY, WEATHER, AND HUMANS

Rocks, as well as landslides, are like people—they don't age well, but rather become weaker over time. Weakness in rocks comes from two main processes: weathering and fracturing. The movement of crustal plates around the surface of the Earth creates regional stresses that permeate rocks down to a microscopic scale. When stresses exceed the strength of the rock, the rocks fracture, just as a stretched rubber band sometimes no longer stretches, but breaks. Over time, weathering by rain, freezing, and thawing also weakens rocks by breaking down the bonds between atoms and molecules. As a result, both microscopic and macroscopic fractures are created.

Landslides are like robbers: they love weakness. Slopes develop a predisposition to failing in landslides because of several processes. At the surface of the Earth, weathering produces a soil overlying the rocks. Vegetation holds the soil together. However, changes in vegetation, either natural or human-caused, can destroy the root systems and weaken the soil. Often there is activity at the base of a slope, such as erosion by a river or waves from an ocean, or land modification by humans. Such activities remove support at the base of a slope. All of these processes predispose a slope to fail and slide when a trigger occurs. By far the most frequent triggers are rainfall and earthquakes.

Nowhere in the US does the effect of rainfall on landslides seem to impact so many as in California. When I was a graduate student at Caltech in the 1960s, we loved to take a spin up the spectacular, cliff-hanging California Highway 39 that ascends through a 30-mile stretch from Azusa to the crest of the San Gabriel Mountains. It was fairly isolated then, but now nearly 3 million people a year drive this highway.[12] In addition to providing a thrill for joyriding tourists, the highway serves residences of 500 people and access to three flood control dams, as well as to the Angeles National Forest for firefighters. But maintenance costs its owner, Caltrans, $1.5 million a year. Here, as everywhere else in southern California, brush fires occur frequently during the dry seasons. These fires destroy the vegetation that holds the soils together. If heavy rainy seasons follow a big fire season, landslides frequently cover the highway. Caltrans, trying to deal with budget problems in 2011, wanted the US Forest Service or Los Angeles County to take over maintenance of the highway. As I write this, both have "gracefully declined to take on this responsibility," said one manager for Caltrans.[13]

The devastating and ongoing consequences of even small landslides are well illustrated by events in La Conchita, a small town covering 28 acres in Ventura County, California. It sits on low land abutting, to the east, a cliff that rises 500–600 feet above the town.[14] At the top of the cliff, a ranch of avocado and citrus trees extends inland for about a half mile. To the west, waves from the ocean lap at the land. It is the perfect setting for a collision of geology, weather, and humans.

In 1995, the ground failed along a fault plane about 100 feet deep. A slow landslide, moving at a few yards per minute, buried nine homes but caused no casualties (Figure 4.2a). However, during a period of heavy rain in 2005, part of the 1995 slide mobilized again, this time moving tens of feet per second, destroying thirteen houses, and killing ten people (Figure 4.2b). The slide buried four blocks of the town with more than 30 feet of earth, a total volume of nearly 2 million cubic yards of material. USGS scientists estimated that the change of state from solid earth to an almost liquid slide occurred "nearly instantaneously."[15]

Townspeople have sued the avocado ranch numerous times, and the detailed story is far too long and complicated to present here. Lawyers for the residents argued that irrigation practices on the ranch caused the slide because irrigation raised the water levels in the hillslopes, weakening them and setting up conditions for the slides. Lawyers for the ranch, however, argued that the irrigation couldn't have been a factor, because there was a history of landslides in this area dating back to at least 1865, when a wagon trail had passed through this part of the coast, well before extensive agricultural watering began.

Southern Pacific Railroad, at great expense, learned of the hazards in this area when it laid tracks there in 1887. Just two years

(a)   (b)

FIGURE 4.2 (a) The 1995 landslide/ debris flow in La Conchita, California. The shallow fault plane where the ground failed is the treeless area at the center-top part of the landslide. (b) The 2005 La Conchita landslide. Note how fluid it appears to have been compared to the 1995 slide. *Part (a): photo by R. L. Schuster, USGS; part (b): photo by Randy Jibson, USGS.*

later, landslides buried sections of the track, and in 1909 slides buried a work train. As a result of these slides, Southern Pacific tried to reduce the hazard by bulldozing flat an area adjacent to the tracks—an act that, unfortunately, inadvertently contributed to future problems. The bulldozing provided prime land on which new construction could occur in dangerous proximity to unstable cliffs. In 1924, two brothers laid out the beginnings of the La Conchita del Mar subdivision.

The lawyers for the ranch also cited a study that showed a strong correlation between intense rainfall and landslides in southern California,[16] and then noted that the La Niña conditions of 1995 had dumped 18 inches of rain on this area, making it the wettest season ever recorded. The 2005 slide also occurred during wet conditions, at the end of a fifteen-day period of near-record rainfall.

After years of legal battles back and forth, the assets of the ranch plus a cash settlement of $5 million were turned over to the plaintiffs. In an interesting political and ethical decision, Ventura County rejected pleas of home owners to stabilize the hillside, arguing that no one should be living in this unsafe area and that any intervention would put the county at risk of future lawsuits. Subsequent research has shown that the 1995 and 2005 slides are part of a much larger prehistoric slide that includes the ranch on the bluffs. Future slides are inevitable.

Although rains are a frequent trigger of landslides, sometimes there is no obvious immediate trigger. The lack of a clear trigger is most common with larger landslides. An earthquake in 1994 shifted a mass of land near the village of Attabad in northern Pakistan, and earthquakes over the next fifteen years shook the region, causing minor damage but no major landslides. Then, without warning, in January 2010 a landslide of nearly 40 million cubic meters in volume struck,[17] damming the Hunza River, killing fourteen people, stranding about 25,000, and inundating 12 miles of the Karakoram Highway, a major trade route between Pakistan and China.[18] Later that year, months of heavy rainfall triggered more flooding and landslides, affecting over 20 million people, killing 2,000, injuring 3,000, and causing $20 billion in damages. The impact on the area was severe, and government compensation was inadequate. Two years after this event, people were still living in tents, at risk of freezing in the cold mountainous area. At one gathering of several thousand people demanding action, serious riots broke out, resulting in two deaths when police opened fire on the crowds.[19]

The situation was even more dire in China in 2010, when floods created especially destructive mudslides there. One such slide, the August 8 Gansu mudslide, was triggered when a debris dam that blocked a small river failed. A five-story-high muddy flow roared through downstream villages, killing more than 1,500 people. Floods and landslides in China that year affected more than 230 million, with over 15 million being evacuated. Even when the torrential rains and high rivers abated, conditions failed to improve, because enormous piles of quick mud were left behind, impeding and endangering rescuers.[20] The UN secretary-general pointed out that the number of people directly affected by this flooding in China exceeded the entire population hit by the Indian Ocean tsunami, the Kashmir earthquake, Cyclone Nargis, and the earthquake in Haiti—combined!

GO . . . NO GO . . . GO . . . NO GO . . . GONE

One of the very few slides that have been observed visually and monitored instrumentally was a giant slide off Mount Toc, Italy, into the water reservoir at Vajont in 1963.[21] The observations provide an intriguing glimpse of what we humans can, and cannot, do when we interact with geological processes.

Between 1956 and 1960, Italian engineers constructed an 860-foot-high dam across the narrow Piave valley (Figure 4.3). During construction, it became obvious that this geological setting had the potential for devastating landslides. Imagine that each wall of the Piave valley consists of a deck of cards tilted at an angle so that the surfaces of the cards are parallel to the wall of the valley. The cards represent the layers of rocks—some strong, some weak. Rocks that are parallel to the land surface like the cards in these tilted decks slide easily, especially if there are weak layers like mud or clay.

In the case of the Piave valley, test boreholes in the rocks failed to reveal any weak layers, and seismic surveys suggested that the rock walls were strong. But the geology in that area led several experts to warn that the entire side of Mount Toc was unstable. In spite of warnings, as part of the postwar reconstruction in Italy the dam was built to provide hydroelectric power, and the results of borehole and seismic studies were used as support for ignoring the warnings.

In 1960, managers began filling the reservoir behind the dam. Although there were minor landslides and earth movements, all went well until the water depth in the reservoir reached about 560 feet. Then, a crack more than a mile long opened across Mount Toc, and shortly thereafter a cube of rock equivalent to the size of a football field on each side (900,000 cubic yards) slid into the lake. Managers of the dam quickly lowered the water level in the reservoir by about 200 feet and the activity stopped. However, it was clear that earth movements and landslides were a problem that had to be dealt with. The engineers inferred that the weight of rising water behind the dam had increased the pressure of water in pores in the rocks, causing the rocks to fracture. (This process is a geological analogy of the engineering practice of "fracking" that is increasingly in the news in the US. Developers of natural gas supplies want to unleash gas stored in rocks by injecting water and manipulating its pressure to create new fractures through which gas can easily flow to wells.)

Because the act of filling the reservoir caused land movement, the engineers felt that lowering the water level might stop the movement. This practice of altering geological processes with engineering design is known as "geoengineering." It is a human practice that in some cases has been successful but more than once has led to trouble. For three years the managers of the

Vajont Dam alternately filled and lowered the reservoir to initiate and stop the movement of the landslide. The geoengineering
practice appeared to work—for a while.

Then, during a rainy period in October 1963, a huge piece of
the mountainside collapsed. A chunk of rock equivalent to seven
football fields on each side (350 million cubic yards) roared
down into the valley. As its potential energy was converted to
kinetic energy during the descent, it reached a speed well over
60 miles per hour. The slide was going so fast that its momentum took it to the height of a forty-six-story building (460 feet)
on the opposite side of the valley. During travel up that side of
the valley, the kinetic energy was converted back to potential
energy and the material came to rest—briefly. Unable to resist
the force of gravity, it then reversed its path back down toward
the reservoir.

The landslide alone would have been catastrophic enough,
but it happened at a time when the reservoir held a substantial
amount of water. The landslide set this water sloshing back and
forth in the reservoir upstream of the dam. The slosh was so high
that it destroyed the village of Casso, which was at the elevation
of an eighty-six-story building above the reservoir level. One can
only imagine the horror of people at the lake who would have had
to sprint eighty-six stories up to escape this wave.

After causing this destruction, a wall of water eighty stories
high then swept over the dam and roared down into the valley.
There it destroyed five downstream villages and killed about
2,500 people. The dam, amazingly, remained intact (see Figure
4.3). Investigations after the event revealed that a weak zone containing small (1- to 6-inch) clay beds within the limestone bed-
rock at depths between 300 and 650 feet had failed.[22]

How is it that a landslide can occur at any time, rather than

FIGURE 4.3 The still-standing Vajont Dam as viewed from the village of Longarone in 2005, showing approximately the top 200 feet. The water that overtopped the dam by 800 feet would have obliterated the sky in this photo. *Photo by Emanuele Paolini.*

just when there is rain or an earthquake? All rocks are riddled with microscopic flaws and fractures. Stresses in rocks are concentrated around the tiny flaws and at the tips of the microscopic fractures, like the weight of a woman is concentrated on the tip of a pointy high-heeled shoe. Stresses are always changing on rocks and slopes, not just when there is rain or an earthquake. For example, the normal tides in the oceans caused by the changing positions of Earth, the sun, and the moon also occur in the solid earth, giving rise to continuously changing stresses. Natural or human-caused variations of water levels in reservoirs, such as those behind a dam or natural lakes, change the distribution of forces on a slope. When a new stress first appears, the old imperfections absorb some of the stress and new isolated microscopic fractures might form. At first nothing dramatic happens, because the old and new microcracks are isolated from each other. But when the overall stress exceeds a critical value, local stresses at the flaws become large enough to break molecular bonds, and the microcracks propagate. The cracks take off running, so to speak, merging to form an ever-larger network of cracks. Ultimately, and usually catastrophically, they grow into the huge surface along which a landslide moves.

In the case of Mount Toc, this process occurred within the weak clay layers that were later discovered to be present deep in the rocks. Water seeping into cracks in the rocks as the reservoir was filled probably played a critical role in reducing the stress that a dry rock could sustain. Every crack ends at a tip that is embedded in atoms and molecules. Water attacks these molecules, giving the crack tip a chemically assisted boost to expand through the rock.

The distance that a landslide travels is called its "runout." Big, fast slides, such as the Elm and Vojant slides, have been given the descriptive name "sturzstrom," a German word for "fall stream" or "collapse stream."[23] A characteristic of these slides is that they travel much farther than would be allowed by normal friction models of the interaction of the landslide with the ground it travels over. It appears that the friction is anomalously low—a puzzle that has perplexed geologists for a very long time.

STURZSTROMS AND FLYING CARPETS

Approximately 17,000 years ago, a volume of rock equal to a cube about a half mile on a side (10 billion cubic feet) roared out of a steep canyon in the San Bernardino Mountains in southern California (Figure 4.4).[24] It originated 1,500 feet above the canyon bot-

FIGURE 4.4 The Blackhawk landslide. *Photo by Michael Collier.*

tom. Rocks in the slide, already fractured at the start of the event, shattered on impact with the canyon bottom, forming intricate three-dimensional jigsaw puzzles. (I shudder to think that this could have been the fate of my neck when I jumped off that roof when I was a kid!)

When this slide, known as the Blackhawk slide, exited from the canyon, it ran out across a nearly flat valley floor for 5 miles. Amazingly, the pieces of the jigsaw puzzles stayed together as the slide zoomed along at nearly 75 miles an hour.[25] A similar landslide triggered by the 1964 Alaska earthquake traveled 3 miles across the nearly level Sherman Glacier before coming to rest[26] (Figure 4.5). Where the base of the landslide could be seen on the glacier, it rested on—believe it or not—undisturbed snow. In other places it left alders, mosses, and small plants completely

FIGURE 4.5 The Sherman landslide on top of the Sherman Glacier. Debris fell from a high peak at upper right. *Photo by Austin Post, USGS.*

undisturbed. The slide climbed over a 460-foot hill, from which we can calculate—from the conversion of potential energy to kinetic energy—that its speed was at least 115 miles per hour.

How can badly shattered rocks maintain their identities through such a long and speedy journey? How can slides leave snow or delicate plants undisturbed by their passage?

The observations of the flying carpet at Elm (see Figure 4.1) and the geological evidence from the Blackhawk and Sherman landslides (Figures 4.4 and 4.5) led to the intriguing hypothesis that landslides can be transported like flexible sheets over a cushion of trapped and compressed air—literally like a flying carpet.[27] Imagine such huge masses of rock roaring down a canyon, hitting a resistant ledge, being launched hundreds of feet into the air, settling back onto a blanket of compressed air only a few feet thick, and then hurling out at breakneck speeds onto the surrounding ice sheet (Sherman) or desert floor (Blackhawk) until the air leaks out and the slide gently glides to a halt—the whole event taking perhaps a minute or two.

In this "air lubrication" hypothesis, the landslide floats as a nearly rigid slab on its cushion of air, so fragile jigsaw pieces of rock like those observed at Blackhawk are preserved. And in this hypothesis, the Sherman landslide can float out over the snowfield of the glacier without scraping or melting the fresh snow underneath, because the cushion of air protects the snow and plants.

Although the air lubrication hypothesis may work for the specific slides for which it was proposed (Blackhawk and Sherman), two arguments suggest that it cannot explain all long runouts. First is the question of whether the air can stay trapped under the slide long enough for the runout, since it would tend to diffuse through the slide and around its edges. The second problem, not obvious when the theory was first proposed, but a result of the observations from unmanned spacecraft looking at the planets and their moons in the solar system, is that long-runout landslides occur on our own moon and on four other moons in our solar system that have no atmospheres—Io, Callisto, Phobos, and Iapetus. Such slides also occur on Mars, which currently has only a very thin atmosphere, although it is not known what the atmosphere was like when the landslides formed in the past (Figure 4.6). Long-runout slides on airless or nearly airless bodies have forced geologists to look at explanations other than air lubrication.

One group of theories takes into account the fact that landslides are not monolithic slabs of rock, but consist of rock fragments of many different sizes.[28] They fall into the broad category of materials called "granular matter" that have fairly unique properties. The cereal in your breakfast bowl provides an example of these properties. Sometimes these materials behave very much like a solid, and other times they flow like a liquid.[29] Grains

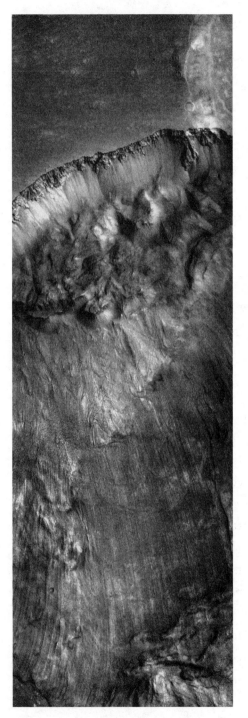

FIGURE 4.6 A large landslide in Ganges Chasma, Mars. The landslide dropped down off a steep scarp, taking out part of the rim of a large, older impact crater (*top right*). The scar is 16 miles wide at the top and 30 miles wide farther down, and it extends 38 miles to its most distant tip. The volume is at least 2,500 cubic miles, five to ten times greater than the largest terrestrial landslides, and the material covers about 800 square miles of the floor of Ganges Chasma to a depth of more than 150 feet. The material dropped at least 6,600 feet from the top of the scar to the bottom of the trough. Los Angeles and Dallas could both fit onto this part of the slide. *NASA* Viking *image.*

can flow, slosh, and reflect from boundaries like a liquid, they can erode channels just like flowing water, and in some instances they can produce hills and gullies that mimic features formed by flowing water.

Two granular-flow theories for sturzstroms require no gas in the atmosphere or in pores in the landslide itself. The premise of one set of granular-flow theories is that the landslide rides on a thin layer of highly agitated dust nanoparticles.[30] This layer is conceptually rather like a layer of air in which dust particles play the role of air molecules. The premise of a second granular theory, called the "acoustic fluidization model," is that during collapse and flow of rocky debris, collisions between particles in the mass create very large and high-frequency pressure fluctuations within the mass as it flows.[31] (To get a sense of this effect, imagine being in your quiet car when a car with a booming, pulsating radio pulls up next to you.) During these fluctuations, the local pressure oscillates between less-than-normal and greater-than-normal. During periods of less-than-normal pressure, stress between particles is reduced and movement can occur.

Increasingly, evidence has been mounting that lubrication is enhanced by liquid water, ice, wet debris, or mud at the base of the slide, or perhaps water within the slide.[32] Even if landslides are not saturated with water, they are unlikely to be completely dry; they will always contain some liquid water (on Earth) or ice (on Earth and the other planets or satellites). As we have seen in earlier discussions, water is very effective at lubricating debris and mudflows. Sometimes water on the surface of the earth even provides a layer over which a landslide hydroplanes like a boat.[33]

Even these possibilities do not exhaust the ideas proposed for long-runout slides. Additional explanations are suggested by the geology of an ancient landslide in Wyoming, the largest known landslide on Earth.

THE MOTHER OF ALL LANDSLIDES:
HEART MOUNTAIN

One of the most fundamental principles of geology states that when sediments are deposited by water or wind, the youngest sediments will be on top. When they are not, a geological mystery awaits. This is the case at Heart Mountain, Wyoming, which sits out in the floor of a basin known as the Bighorn Basin. The base of Heart Mountain consists of rocks that are only about 55 million years old, but at the top the rocks are 350–500 million years old. How did this inversion happen? It's been a puzzle to geologists for over 100 years.

By careful geological mapping, geologists have figured out that the rocks now on top of Heart Mountain were at least 30 miles away from the mountain 50 million years ago, over in the Beartooth Mountains to the northwest (see Figure 4.7). During an episode of geological mountain building about that time, the Beartooths rose up and tilted toward the east like a stack of dinner plates sliding off the tray of a waiter in a restaurant. Immediately (by geological standards) after this uplift, volcanic eruptions began in the Absaroka Range, to the south of the uplifted area. A giant sheet of the old rocks, perhaps 500 square miles in area and 2 miles thick, slid down from the Beartooth Mountains toward the Bighorn Basin. This massive landslide traveled south and southeast, up over a 1,600-foot-high mountain, and finally came to rest on the younger rocks of the Bighorn Basin, and on top of Heart Mountain. This landslide flowed more than 30 miles over terrain with only a 2-degree slope, spreading out to cover 1,300 square miles. The whole event probably lasted less than an hour.

Geological evidence suggests two culprits involved in the initiation and movement of the Heart Mountain slide: the 50-million-

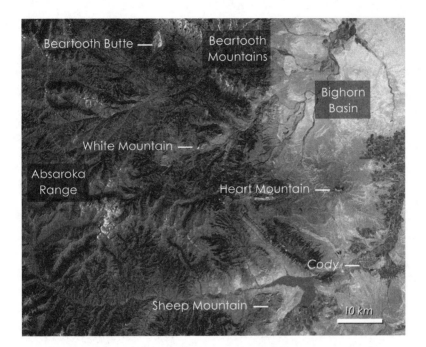

FIGURE 4.7 Heart Mountain. *NASA image by Robert Simmon, based on data from Landsat 7.*

year-old volcanic rocks of the Absarokas, and the carbonate rocks of the Madison Limestone. But with these two potential culprits, geologists have proposed more than a dozen models. One possibility is that volcanic eruptions in the Absarokas triggered the landslide. Another is that heat from magma rising under the old rocks raised the pressure of water in the rocks, which weakened them and triggered the movement. A third is that the hot volcanic rocks transformed water surrounding the volcanic system into steam that then lubricated the fault surface. A fourth possibility is that once the landslide started, pressures during its motion melted the rocks, forming pseudotachylite that lubricated the faults as has been proposed for its role in earthquakes (Chapter 3). A fifth is that pseudotachylite that had formed from limestone (calcium carbonate) in the rocks broke down chemically to produce carbon dioxide gas that provided the air-like lubrication over which the slide hovered as it moved.[34] A recent model examining this last possibility led to the proposal that the slide moved at velocities between 300 and 700 miles per hour and that the whole event lasted only three to four minutes.[35] Whatever the details of the motion of such huge landslides, it is clear that they are capable of massive destruction in a very short time.

REFLECTIONS: GEOLOGY, A BIFOCAL SCIENCE

In the Reflections section of Chapter 3, I discussed "rare" events and some of the problems they pose for risk communication. Here I would like to explore further the meaning of the word "rare."[35]

Geologists use the word "rare" in the context of both space ("spatially rare") and time ("temporally rare"). In the context of space, rare means "widely spaced," and in the context of time, it means "occurring infrequently." In both cases, the word is synonymous with "uncommon."

Some phenomena are of such a nature or scale that they are truly uncommon in any sense—the largest landslides on earth and in the solar system discussed in this chapter are examples. But, rareness also depends to some degree on the perceptions of the observer. People who have few opportunities to travel may see for the first time a new landscape and think that it is rare because they haven't encountered it before. For example, I grew up in northwestern Pennsylvania, a region of forested rolling hills, in a home where we had no television, nearly a half century before the global availability of information and images through the World Wide Web. Although I read a lot and devoured *National Geographic*s, when I first saw the deserts of the American Southwest at age twenty-two, I was totally unprepared for its vast landscape and exotic cacti. To me, at that time in my life, a desert seemed rare. Yet after living in the Southwest for many years and becoming a geologist, I learned that deserts were not uncommon, that this impression had been only a limitation of my perception. In these days of easy access to information on the Internet, most people can, if they wish, discover much about exotic locations far away from their homes even if they do not do much traveling.

Perception of events that are rare in time is, however, more difficult. We simply don't focus on phenomena that are not happening in the present, or near present. As discussed in Reflections in Chapter 3, statistics about the frequency of rare hazards don't preclude our living in areas subject to them. This is probably the main reason that so many people live in places exposed to the large geologic hazards that occur infrequently—large earthquakes, megalandslides, huge volcanic eruptions, or big tsunamis.

Geologists tend to be world travelers, and when we gather together on field trips, you often overhear comments such as "Oh, this feature in Asia reminds me of that one in the United States" or "This is amazingly like something I saw before except . . ." or "This event appears not to have happened in the past few millennia, but I think that the rocks in such-and-such a place tell us that this has happened before." Geologists also tend to be the "elders" of the planet because we study events that happened as far back as billions of years in the history of our planet. We look at the world metaphorically both through a telescope and through a microscope.

I have called geology a "bifocal science," from the analogy to bifocal lenses that allow people with vision problems to see both near and far objects. The bifocal perspective gives geologists the ability and perspective to focus on the possible occurrence and form of events in our nearby space and at great distances, and in the past, present, and future times. The landslides mentioned in this chapter range in age from nearly contemporaneous to tens of millions of years or even, in the case of Mars, billions of years. They range in scale from small to enormous and occur all around the world and on other planets.

By studying landslides at these many scales, geoscientists have come up with such a bewildering array of proposals for how they move that it may sound as if we simply don't know what we are talking about. Some of the processes proposed almost certainly occur some of the time in some of the landslides. Not all of the processes occur all of the time or in all places. The large number of hypotheses and mechanisms proposed is simply testimony to the awesome complexity of our world, not to our ignorance. To add to this complexity, in the next chapter we will see that one geological disaster, such as a volcanic eruption, can initiate a chain reaction that unleashes other disasters.

Chapter 5

THE DAY
THE
MOUNTAIN
BLEW

BLA-LOOM!

"BLA-LOOM! Like hell! It is the mountain. The mountain is doing something."[1] This is how six-year-old native Alaskan Harry Kaiakokonok described the 1912 eruption of Novarupta in Alaska—the largest eruption of the twentieth century—in which 3 cubic miles of magma erupted, destroying the area now known as the Valley of Ten Thousand Smokes.

In 1980, David Johnston, a young geologist with the US Geological Survey, holding a lonely vigil on a ridge about 6 miles northeast of Mount St. Helens, Washington, also witnessed an eruption. Six weeks earlier, the volcano had emerged from a slumber of 123 years with seismic rumblings underground and the emission of small, ash-laden plumes from its summit. Because of its youth and presence in the active volcanic chain of the Cascade Range, the volcano had been under constant observation since its awakening. The weather on the morning of May 18 was crystal clear, and from his vantage point Johnston saw a large landslide falling off the north flank of the mountain, and a massive eruption blasting forth toward him. He radioed to headquarters, "Vancouver! Vancouver! This is it!" He was never heard from again.

Nearly twenty centuries earlier, in AD 79, Pliny the Younger, at eighteen years old, observed the eruption of Mount Vesuvius, Italy, from a distance that permitted more observation. This massive eruption destroyed three cities—Pompeii, Herculaneum, and Stabiae—killing an estimated 16,000 people, including the young man's uncle, the famous naturalist and philosopher Pliny the Elder. After his mother called his attention to an unusual cloud, Pliny the Younger wrote down his observations:[2]

> Its general appearance can be best expressed as being like an umbrella pine [Figure 5.1], for it rose to a great height on a sort of trunk and then split off into branches, I imagine because it was thrust upwards by the first blast and then left unsupported as the pressure subsided, or else it was borne down by its own weight so that it spread out and gradually dispersed. Sometimes it looked white, sometimes blotched and dirty, according to the amount of soil and ashes it carried with it.

Vesuvius is still an active volcano, now looming above 3 million people in Naples and surrounding regions (see Figure 1.1). Only recently, geologists discovered evidence of an even larger blast by Vesuvius about 3,800 years ago.[3] This eruption laid down 15 feet of ash in what is now the center of present-day Naples. Seismic evidence shows that presently a layer of anomalous properties about 6 miles underground could be volcanic magma (but arguably it could be water or brine). While scientists debate the geology and probabilities of eruptions of varying style and intensity at Vesuvius, emergency response planners debate not only how to evacuate so many people, but how even to decide whether or not to evacuate.[4]

(a) (b)

FIGURE 5.1 *(a)* The Plinian column over Vesuvius, illustrating the "trunk" of an eruption column and its "branches," or "umbrella." As seen from Naples in October 1822. *(b)* An Italian stone pine of the type described by Pliny. *Part (a): drawing by George J. P. Scrope, 1864; part (b): image provided by Jose Cao-Garcia.*

A disproportionate fraction of our civilization is exposed to the effects of eruptions, including major centers of population and commerce such as Seattle, Tokyo, Mexico City, Rome, Manila, Auckland, and Quito. Popocatépetl looms over Mexico City and its 19 million residents. In 2000, the government evacuated tens of thousands of people prior to one of its biggest eruptions in 1,200 years. In April 2012, activity in "Popo" increased sufficiently to cause some schools to be closed, and to prompt officials to alert residents to close their windows and avoid being outdoors because of the fumes. Popo has had fifteen fairly major eruptions since the arrival of the Spanish in the sixteenth century, generally spewing ash and gas, but it is capable of unleashing much greater devastation.

The volcano Nevado del Ruiz looms over the town of Armero, Colombia, having a history of only a few relatively small eruptions, but also a history of producing massive hot mudflows, known as "lahars." In 1985, 22,000 people, three-quarters of the population of Armero, died when lahars swept down valleys on the volcano's flank (Figure 5.2). During this eruption, hot ash that spewed from a vent at the summit landed on the snow, ice, and glaciers, melting them and lubricating the lahars, which then traveled down the valleys at speeds averaging about 25 miles an hour. Flows like this are a likely consequence of a future eruption of our own Mount Rainier, Washington, where more than 150,000 people live on top of the deposits left by past flows.

In 2009, much to the chagrin of volcanologists, in a widely televised response to President Obama's presidential address, Governor Bobby Jindal of Louisiana mocked the volcanic monitoring program of the US Geological Survey as "wasteful spending" and proposed its elimination. Much to the relief of volcanologists and of people in the Pacific Northwest, Alaska, and Hawaii who live near the volcanoes in the US, the eruption of Eyjafjallajökull[5] ("Eyja") in Iceland the following year, and its huge global economic impact, laid that shortsighted suggestion at least temporarily to rest.

All of the geological processes discussed to this point—earthquakes, landslides, tsunamis—are terrifying for those who experience their full fury. But a massive volcanic eruption dwarfs any other and has the potential to inflict a final blow on civilization. Although volcanoes occupy only a small area of the planet, their influence far exceeds their relative area because these areas include some of the most hospitable and fertile lands on the planet. At one point in the late twentieth century, the US Geological Survey

FIGURE 5.2 Hot lahars swept down the valley and out onto the town of Armero, Colombia, in 1985, burying much of it under mud. *Photo by R. J. Janda, USGS.*

estimated that volcanoes posed a tangible risk to at least 500 million people around the world.

The historic and geological records are very clear: volcanic eruptions come in various sizes and styles. The intervals between eruptions can be long or short, but eruptions are inevitable. Just the time and place remain to be determined. Gigantic eruptions make great fodder for Hollywood productions, but in fact, the more common smaller and intermediate-scale eruptions are sufficient to cause massive disruption.

Volcanic eruptions kill people in four ways: by flows of hot ash and gas, by flows of hot mud, by tsunamis, and by famine and disease. In the last 200 years, 200,000 people have died from volcanic eruptions, about one-third of them from starvation and disease. More than 9,000 Icelanders died of starvation in 1783 from the toxic mixture of gases and ash that also decimated their livestock when Laki erupted. The 1815 eruption of Tambora in Indonesia resulted in 92,000 deaths, mostly from starvation. A tsunami caused by the 1741 eruption of Oshima in Japan killed most of the nearly 1,500 victims of that eruption. The tsunami caused by the 1883 eruption of Krakatoa was responsible for 36,000 deaths.

What determines the power of volcanic eruptions? How is this power measured? How high do eruptions go, and what determines the height? To address these questions, we'll visit Mount St. Helens and Mount Pinatubo in the Philippines on our field trip before taking a look at a few gigantic eruptions in the not-so-distant past. In these visits we'll see evidence of one of the most powerful changes of state in nature: the production of gas in volcanic systems.

BREWING UP A DANGEROUS MIX

Volcanic eruptions come in an amazingly wide variety of manifestations: some send soaring plumes tens of thousands of feet into the atmosphere, whereas others exude ground-hugging, dense, fiery "ash hurricanes." Excluding the interaction of magma with external water and ice, which can make even the most benign eruption explosive, the character of an eruption depends on the composition of the magma.

Volcanic magma is proof that a very complex material can be formed out of some simple building blocks—about a dozen elements from the periodic chart and a few gases. The physical properties of magma change in a systematic way with the proportions of these elements. The major ones are silicon, oxygen, and aluminum, with lesser amounts of iron, magnesium, calcium, potassium, and sodium. Depending on the proportions of these elements, there are many types of magma, but it's sufficient to know about the four most common, which you can remember by using as a mnemonic the Shakespearean term "BARD," for basalt, andesite, rhyolite, and dacite. Basalts have the least silicon and oxygen, and the most iron, magnesium, and calcium. They are very fluid (that is, their viscosity is low), and they don't contain a lot of dissolved gas. They melt and erupt at the highest temperatures of these four, typically producing red-hot lava flows or mildly spattering, gently explosive eruptions. The tourist-friendly effusions of lava in Hawaii are basaltic.

At the other end of the BARD spectrum, rhyolites and dacites contain the highest amount of silicon and oxygen, are generally very viscous, contain the most dissolved gas, and melt and erupt at the lowest temperatures. They can produce either very viscous lava domes, or very explosive eruptions. The explosive eruptions discussed in this chapter, such as those of Mount St. Helens and Mount Pinatubo, were driven by dacites. Andesites, often formed by the mixing of basalts and rhyolites or dacites, are intermediate between basalts and rhyolites or dacites in all of these properties.

The explosivity of these eruptions comes not from the melted rock, but rather from the gases that are dissolved in it—typically water vapor (H_2O), but also carbon dioxide (CO_2) and sulfur dioxide (SO_2). The magmas that originate from the melting of the subducted slabs of oceanic crust around the Pacific plate in the so-called Ring of Fire (see Figure 3.1) are typically andesites and dacites, and can contain up to 5 or 6 percent water. That's potentially a lot of gas (steam); for comparison, a typical soda pop, which seems very gassy when shaken, has only about a half percent of gas. These are the magmas that can produce big disasters. Their devastating potential is often increased by the involvement of groundwater in the volcano, or in glaciers and snow on its slopes.

Magma can hold more gas in solution at high pressure than at low pressure, so as magma rises up in a volcanic system, gases come out of solution, forming small bubbles. Initially, as the bubbles form, magma looks like the stuff in a soda bottle that is shaken with its top still on: little bubbles are scattered throughout the liquid. As the magma rises to lower pressures, the texture takes on the characteristics of foam, and the bubbles in the foam get bigger and bigger as it rises higher and higher.

Under some conditions, the rising magma is relatively cool and viscous. Unable to migrate toward the surface, the gas bubbles become trapped in the magma. In such instances a viscous plug rises slowly in the conduit. It might break through to the surface, forming a viscous dome, as was the case at Goat Rocks, a bump on the north flank of Mount St. Helens formed in the mid-1800s.

In other situations, the gas bubbles can expand to the point that the walls containing them break, freeing the gas and loading it with shards of molten material that solidifies to become ash. This typically happens when the bubbles reach about 75 percent of the total volume. When the bubbles burst, internal energy that was stored in the compressed, pressurized bubbles transforms to kinetic energy. If the resulting mixture of gas and ash has a high density, it pours down the slopes of a volcano in the form of deadly ash hurricanes; if it is less dense, it rises into a plume that towers over the volcano.

The structure of the underground reservoirs that house magma before eruption has puzzled volcanologists for a long time. These reservoirs are generally tens of miles underground, so the magma has to travel a long way to erupt. Yet volcanoes are clearly capable of expelling a lot of magma in a very short time. In the two largest volcanic eruptions of the twentieth century, Novarupta in

Alaska erupted 3–3.5 cubic miles of magma in an estimated 60 hours in 1912, and Pinatubo (Philippines) erupted a bit over 2.5 cubic miles in just a few hours in 1991.[6]

Cartoons, even in some textbooks, of the structure of volcanoes often show a balloon-shaped cavern containing a liquid magma that is ready to squirt out under high pressure. That may look like a configuration that could erupt a lot of magma fast, but geophysical studies reveal that there is a problem with this simple picture. Shear waves generated by earthquakes should vanish if they traverse a volcanic region with big reservoirs of liquid magma—but they do not.

To account for the seismic data, volcanologists have proposed that a volcanic reservoir is filled with a "magma mush." The mush is an intricate mixture—lenses of pure melt interspersed with "mushy melt" (a material that is only partly melted, like a partially defrosted stew), or even nearly solidified magma in contact with solid, mildly warm rock at the edge of the reservoir. Such a substance would have enough solid material that it could support shear stresses and transmit shear waves, and yet enough liquid that it could become mobile during an eruption. Magmas that underlie the volcanically active midocean ridges may consist of 80–90 percent mush and 10–20 percent pure liquid melt.[7] Seismic evidence suggests that the hot rocks underlying Yellowstone contain 10–30 percent melt, and one argument that an eruption of Yellowstone isn't an imminent threat is that this is too little melt to drive an explosive eruption.

How does a dense, often viscous, magma mush at depth become a hot, gassy ash cloud roaring out of a vent at the surface of the earth? We begin to answer this question by looking at magma that almost didn't make it: the magma rising in Mount St. Helens in 1980.

THE CALM BEFORE THE STORM
AT MOUNT ST. HELENS

Geological studies of old eroded volcanoes have shown that most magma never reaches the surface to erupt into the atmosphere. It is generally stored along the way in pockets and layers that are revealed only when erosion has exposed the insides of the volcanic plumbing system. However, geological events can sometimes disrupt this course of events. We found this out the hard way in 1980, when magma began rising into the summit of Mount St. Helens.

On March 20 that year, a magnitude 4.2 earthquake at the mountain alerted geologists that volcanic activity might be imminent. A week later, steam started venting vigorously through the summit. Was the volcano just burping, or was a major eruption materializing? A small cadre of scientists responded to monitor the situation by showing up in Vancouver, Washington, where an improvised center had been established in a US Forest Service building. Some were veteran volcanologists with experience at volcano observatories and in research work in Hawaii and at the Cascade volcanoes. I was not such a veteran; I had no experience with volcanoes and had been studying geysers in Yellowstone National Park with the idea that their eruptions might help me learn about volcanic eruptions. I showed up with only my enthusiasm and a trusty Super 8mm camera that I had been using to film the geyser eruptions. I believed that by analyzing films of the eruptions, I could understand conditions underground, and by comparing volcanic eruptions with geyser eruptions, I could learn about the thermo- and fluid dynamics.

On a snowy, gray, cloudy, and miserably cold day a week after the volcano started erupting, two of us reached the end of a snow-covered logging road north of the mountain. The mountain itself wasn't visible anywhere through the snow and fog, so, using a compass and a map (this was before the days of GPS), we oriented our small tent so that we might open the flap with a view of the volcano—should the weather ever clear. The uncomfortable night was spent trying to keep both ourselves and our cameras warm inside sleeping bags.

At dawn, we were awakened by thunder. This didn't make sense. It wasn't raining, and the sun was shining brightly through the thin fabric of the tent. We opened the tent flap to a view of a magnificent eruption, complete with thunder and lightning (Figure 5.3). Yep, we had managed to line up the tent right, and Mount St. Helens had awakened from its repose. Active volcanoes in the US were no longer just picturesque mountains in Hawaii and Alaska, but one was erupting in our own backyard in the contiguous states.

Mount St. Helens is but one volcano in the Cascades, a range that extends through the northwestern part of the US into southern British Columbia, Canada. Here, the rocky Juan de Fuca Plate, covered with wet sediments from the ocean bottom, is being subducted below the continental crust of the North American Plate. As the plate and these sediments are pulled down, the pressure and temperature on them increase, causing them to melt and form magma. The magma, being less dense than the crust of the North American Plate, works its way up through the crust to supply the active volcanoes—Rainier, Baker, and Adams, to name a few. Mount St. Helens is the youngest, and one of the smallest, in this chain. It has a history of eruptions extending back about 37,000 years. Oral and written records attest that the most recent eruption cycle before 1980 had started in 1800 and continued intermittently for 57 years until the volcano went back to sleep.

FIGURE 5.3 A typical small eruption from the summit of Mount St. Helens during March 1980. *Photo by S. Kieffer.*

The eruptions of March and April 1980 threw rock and ash that had been deposited in the edifice by old eruptions out of the summit, creating a small crater in the top of the mountain. The crater grew until mid-April, when the eruptions quieted down. As would be the case for Eyja thirty years later, small eruptions were an indication that something bigger might be in the works. The difference was that the early eruptions of Eyja spewed out fresh magma readily available in its Icelandic setting, whereas at Mount St. Helens, the magma was struggling up from depth as a viscous plug but had not yet appeared. It was, however, close enough, and hot enough, to heat water in the mountain to drive the steam eruptions. They were geyser eruptions[8] rather like those that tourists see every day in Yellowstone—just larger and dirtier.

THE STORM: AN ASH HURRICANE

Ominously, during this time, the north side of the mountain began to bulge in the vicinity of a rock feature known as Goat Rocks, the site of eruptions in the 1800s. Seismic monitoring and measurement of the shape of the north flank of the volcano showed that it was bulging out at speeds of up to 3 feet a day as magma pushed its way up from depth. The bulging edifice became more and more unstable until, at 8:32 a.m. on May 18—a time and date forever etched in the minds and emotions of the geologists who worked there—the weight and stresses of the bulge could no longer be sustained. The north side just fell away in a large and complex series of landslides.

The landslides created a large amphitheater-shaped gouge about a half mile wide in the north flank of the mountain. The removal of this mass of material nearly instantaneously reduced the pressure on the buried magma plug, releasing a lethal mixture of gas, ash, and rocks. The leading edge of a huge dark gray cloud spewed north out from the amphitheater across the land at nearly 225 miles per hour in an event now referred to as the "lateral blast" because it was well described as a blast, and was so laterally focused. Models of the blast suggest that velocities inside the blast cloud could have reached 1,000 miles per hour. The blast devastated more than 230 square miles of forest, stripped trees and soil from the land, and killed all life, including fifty-seven people, among them David Johnston, the USGS geologist on duty at the mountain.

Ash-laden flows of hot volcanic gas have been referred to as "ash hurricanes." Close to the mountain, the trees and the soil supporting them were completely stripped from the land. Farther out, the trees were sandblasted to shreds, leaving only stumps. Farther out, many of the trees, missing all their large and small limbs, were flattened to the ground like matchsticks pointing

FIGURE 5.4 Car of a *National Geographic* photographer, Reid Blackburn, who was killed by the Mount St. Helens blast. The car is partially buried in powdered rock and ash from the May 18, 1980, eruption. *Photo by Dan Dzurisin, USGS.*

away from the volcano. Their root balls pointed back toward the volcano, but their roots—no longer in the ground—were wrapped clear back around the root balls by the force of the ash hurricane. Heavy logging equipment was destroyed, some being overturned like toy Tonka trucks. Even cars provided no shelter for those caught in the area (Figure 5.4).

The scoured earth and downed trees preserved a detailed, and puzzling, record of the direction of the blast as it traveled across the land (Figure 5.5). Many ash hurricanes follow the contours of the land—a behavior apparently first described by the French geologist Antoine Lacroix in the 1902 eruption of Mount Pelée on Martinique in the West Indies.[9] There, in one of the deadliest eruptions of the twentieth century, hot, glowing ash, reaching nearly 2,000°F, inundated the city of Saint-Pierre. The dense clouds of ash poured out of a notch in the summit of the volcano and, driven by gravity, rushed down its flanks, igniting everything in their path along the way. It was from this eruption that the French term *nuée ardente* ("glowing avalanche") came into common use to describe these flows. Like water in a river, the *nuées ardentes* flowed downhill, except where their momentum drove them up and over local obstacles. Over a broad area away from the vent of Mount St. Helens—the outer zone labeled the "channelized blast zone" in Figure 5.5—gravity and momentum drove the flow.

But in some areas the lateral blast at Mount St. Helens defied such simple laws—the inner zone labeled as the "direct blast zone" in Figure 5.5. Over distances up to 8 miles north of the volcano, the tree blowdown revealed a different behavior. Without regard to even thousands of feet of relief between valleys and mountains in the landscape, the trees just pointed straight away from the volcano, indicating that the blast had traveled like that old army song: "over hill, over dale." Something in addition to gravity was driving the flow.[10]

When I looked down from a helicopter at the barren landscape and its devastated forest, it seemed to me that a giant rocket nozzle, lying on its side with the engines pointing north, had discharged from the amphitheater across the landscape. In fact, this was a clue to the additional force beyond gravity driving the blast: the discharge and decompression of a gas from a high-pressure reservoir (that is, inside the mountain) into a lower-pressure reservoir (the atmosphere), like the gases in a discharging rocket or in the bicycle tire blast described in Chapter 2.

How fast is that observed velocity of 225 miles per hour? Nearly as fast as professional race cars and about half the speed of modern commercial aircraft. Another way to make this comparison is to look at the Mach numbers of the blast and an aircraft. Commercial jets travel at a Mach number of about 0.7–0.8,

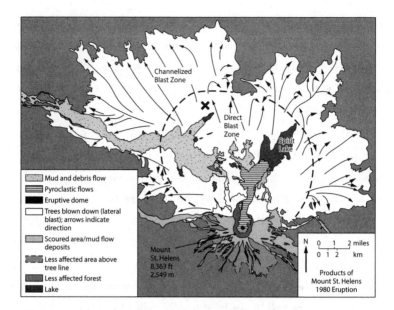

Mud and debris flow
Pyroclastic flows
Eruptive dome
Trees blown down (lateral blast); arrows indicate direction
Scoured area/mud flow deposits
Less affected area above tree line
Less affected forest
Lake

Channelized Blast Zone

Direct Blast Zone

Spirit Lake

Mount St. Helens
8,363 ft
2,549 m

N   0   1   2 miles
     0 1 2   km

Products of
Mount St. Helens
1980 Eruption

FIGURE 5.5 The area north of Mount St. Helens that was devastated by the 1980 lateral blast that killed the USGS geologist David Johnston, whose approximate location is marked with an "X." The tree blowdown directions are generalizations from detailed hand-drawn maps of small areas of the tree blowdown by the author. These are available at http://www.geology.illinois .edu/~skieffer/maps.php.

but according to models of the fluid dynamics, the Mach number of the lateral blast was between 1 and 4.[11] How can it be that material traveling slower than a jet can have a higher Mach number? To understand this dilemma, we have to look at Mach numbers not only in the common context of air travel, but in the way that fluid dynamicists look at them.

The most commonly used definition of the Mach number is the ratio of the velocity of an object, such as an airplane, to the sound speed of the medium in which the object is moving—for example, air. The sound speed for gases depends (inversely) on the molecular weight of the gas; it is very high for the light gases, helium and hydrogen (about 2,200 miles per hour); lower for air (750 miles per hour); and very low for a heavy gases like the now-banned refrigerant Freon (350 miles per hour). This means that, according to the normal calculation, the Mach number would be about 0.3 (225/750) for the lateral blast, roughly half that of the commercial jet and in agreement with normal expectations that a lower velocity gives a lower Mach number.

So why do I say that the blast had a Mach number of 1–4? If any gases contain a lot of rocks, dust, crushed glaciers, and shredded trees, their sound speed is lower, sometimes significantly lower, than the pure gas. Fluid dynamicists use a model called a "dusty gas model" to approximate such complicated mixtures as a gas with a very heavy molecular weight. In accordance with the inverse dependence of the sound speed on molecular weight, the sound speed of such a substance can be significantly less than that of the pure gas, the exact figure depending on how heavily laden the gas is with dust and rock. The best estimates of the sound speed of the blast mixture put it at about 225 miles per hour,

Now we arrive at the second way to view Mach numbers. Fluid dynamicists compare the velocity of a moving gas with its *own* sound speed instead of the sound speed of the medium through which it is moving. At 225 miles per hour, the blast was traveling at about Mach 1 (225/225) in this sense, in contrast to 0.3 (225/750) relative to the external sound speed of air. Detailed calculations showed that Mach numbers reached values between 3 and 4 inside the blast.[12] That is, the plume was traveling subsonically with respect to the atmosphere, but strongly supersonically internally.

The reason this is important is that a high Mach number indicates that the pressure generated by the expanding gas is greater than the force of gravity. Here lies the explanation for the direct blast zone surrounding the vent of the lateral blast at Mount St. Helens. The direction of the downed trees in this zone had little rela-

FIGURE 5.6 Test firing of an F-1 engine of the type that powered the Saturn V rockets that took astronauts to the moon. Supersonic exhaust is visible at the bottom center. *NASA image.*

tion to the hills and dales that should have deflected the flow into valleys. Rather, the trees pointed directly away from the volcano in the direction that the high-velocity internally supersonic flow was traveling. The high velocity of the expanding blast also deeply scoured the landscape, removing trees and soil over a significant area. Only when the gas decelerated to much lower velocities and subsonic speeds did the trees fall in the direction dictated by gravitational forces in the channelized blast zone. According to the model, the high-velocity core of the direct blast zone and lower velocities in the channelized blast zone were separated by a shock wave standing within the blast flow at a distance of a few miles from the vent.

A comparison of the lateral blast with the engines that launched the astronauts to the moon is humbling. A Saturn V rocket is propelled by a set of five F-1 engines (Figure 5.6). The area over which the engines discharge the gas is about 525 square feet—a bit larger than the area between the goal line and the 1-yard marker of a standard professional-league football field. The amphitheater through which the lateral blast poured to the north was about 5,000 times as large. The combined power of the five F-1 engines is about 500,000 megawatts (MW). For comparison, this is slightly more than 100 times the installed capacity (4,500 MW) of the ill-fated Fukushima power plant in northern Japan. The power of the lateral blast was 13,000 times greater than that of a Saturn V, or 1.4 million times that of Fukushima. The thrust of the Saturn V at liftoff is 7.6 million pounds force; that of the Mount St. Helens lateral blast was 100,000 times greater.[13]

UP, UP, AND AWAY

For about three hours after the lateral blast at Mount St. Helens, the volcano spewed forth a dark mixture of rocks, ash, and gas as the

FIGURE 5.7   Plinian eruption column during the May 18, 1980, eruption of Mount St. Helens. *Photo by Austin Post, USGS.*

eruption cleared a throat from the magma to the summit. Around noon that day the character of the eruption changed dramatically with the emission of a gray plume of steam and ash emerging from the summit (Figure 5.7), accompanied by dense flows of hot gas and ash (called "pyroclastic flows") down its flanks. A towering eruption column reaching 13 miles into the stratosphere eventually developed over the summit of the volcano. Although detected by radar, its upper reaches were largely obscured by dust and clouds that had developed after the lateral blast.

A song written by Jimmy Webb and performed by a number of popular groups in the 1960s, called "Up, Up and Away," aptly describes what happens when gas gets loose in a volcanic system. In an eruption column, velocities often reach 900–1,300 miles per hour. Although difficult to measure directly,[14] computer simulations suggest that these flows are internally supersonic for thousands of feet, or more, above the vent. However, just as in the case of the lateral blast, the high-speed core and any internal shock waves are almost always concealed by a sheath of gray, white, and black material that appears to be slowly swirling around its exterior. The high-speed core, sometimes referred to as the "jet thrust region," forms the near-vent region of a volcanic plume. Shock waves within the eruption column play a role in determining the velocity of the volcanic jet and, through this, its height. The height, in turn, determines how far the plume spreads out in the atmosphere and, ultimately, the area of land affected by fallout of ash from the plume.

As the jet travels higher and higher, it gulps in more and more air. This air, cooler and denser than the hot volcanic gases driving the plume, bogs the jet down. This slowdown is, however, resisted by another simultaneous process: the transfer of heat from the hot ash fragments to the entrained air. If there is enough heat transfer, the plume continues its upward journey because it has become buoyant compared to the atmosphere surrounding it. It rises like a toy balloon filled with light helium.

Eventually, the jet loses momentum and buoyancy. It can then exhibit one of two behaviors: It can collapse back toward the ground, forming gravity-driven pyroclastic flows. Or it can stall, but not collapse, at a height referred to as the "neutral buoyancy level." (When you are floating easily at a particular level in fresh- or salt water, depending on your body build, you are at your own neutral buoyancy level.) As long as material is pumped into the base of the eruption column, it will rise up in the column and, when it reaches the top, spread out laterally to form a broad horizontal cap called an "umbrella." This is the process that forms the "trunk and branches of the pine tree" described by Pliny the Younger in his quote at the beginning of the chapter (see Figure 5.1).

Only eleven years after Mount St. Helens erupted, the second-largest eruption of the twentieth century was unleashed when, after 500 years of slumber, Mount Pinatubo in the Philippines came alive in 1991. At least 30,000 people lived in small villages around its flanks, and a half million lived in cities surrounding the volcano. Clark Air Base, a major American base during the Cold War, lay about 15 miles to the east. Fortunately, because of improved monitoring techniques resulting from lessons learned during the 1980 Mount St. Helens eruption, 60,000 people were evacuated from the slopes and valleys. The American military evacuated 18,000 personnel and equipment from the air base. Thousands of lives and a billion dollars in property were saved.

In March and April that year, magma wound its way up to the surface from a depth of more than 20 miles, its presence announced by myriad small earthquakes and, eventually, by steam explosions that created craters on the north flank of the volcano.[15] In those spring months, thousands of tons of foul-smelling sulfur gas were released—to geologists a clear sign of the presence of magma. By the end of the eruption, more than 20 million tons of sulfur dioxide had been ejected into the atmosphere by the eruption.

On June 12, Pinatubo began a sequence of explosive eruptions, producing devastating ash-laden surges that swept down the flanks of the volcano and tall Plinian eruption columns that reached more than 20 miles up into the atmosphere. On June 15,

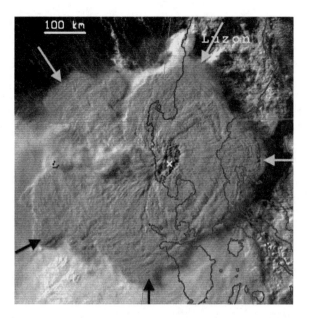

FIGURE 5.8 Satellite photo of the plume from Pinatubo.
The solid black outlines show the islands of the Philippines, the "X" marks the location of Pinatubo, and the
five arrows show the lobes on the umbrella. *From
P. Chakraborty et al., "Volcanic Mesocyclones," Nature
458 (2009): 497–500.*

Typhoon Yunya made landfall only 45 miles from Pinatubo. Perhaps by coincidence, or perhaps not, Pinatubo entered the climac-
tic phase of its eruption that day. Hot, dense flows of ash roared
down the slopes of the volcano, destroying seismometers needed
for monitoring in the bad weather, and turning day to night. Satellite images showed that the Plinian column expanded like a giant
umbrella into the stratosphere (Figure 5.8), reaching 250 miles
in diameter in only two hours. The darkness was compounded,
and the effects of the eruption enhanced, by Yunya. Heavy rainfall created a rain-saturated blanket of muddy ash covering 3,000
square miles of the island.

The volcano ejected 1.3 cubic miles of ash, excavating such a large hole in the earth that the eruption ended with the formation of a crater ("caldera") 1.5 miles in diameter. The collapse of the caldera choked off the supply of magma and shut down the eruption.

From an analysis of four satellite images of the Pinatubo plume, the only ones that existed of its umbrella, scientists documented that the immense umbrella was not circular but instead had five lobes, and that it was rotating.[16] The rate of rotation was not huge—about 10 degrees per hour after two hours—but it was sufficient to provide a dynamic instability that gave what would otherwise have been a circular umbrella its lobate structure when viewed from satellite.

This was the first—and, to date, only—observation of a rotating plume since 1811, when a sea captain described the emergence of a volcanic vent and eruption column out of the ocean in the Azores, a chain of nine volcanic islands in the North Atlantic.[17] The captain described the column rotating "like an horizontal wheel." He also noted that the eruption was accompanied by flashes of lightning that "continually issued from the densest part of the volcano." Then, as he put it, the column "rolled off in large masses of fleecy clouds, gradually expanding themselves in a direction nearly horizontal, and drawing up to them a quantity of waterspouts."

Although some of these features have been described individually in other eruption columns—such as waterspouts at Surtsey, an island off of Iceland (Figure 5.9a) or lightning at Chaitén, a volcanic caldera in Chile (Figure 5.9b)—the sea captain's account appears to be the only one of a volcanic plume in which rotation of the eruption column, lightning, and waterspouts were all occurring simultaneously in a volcanic plume.

The newly discovered rotation phenomenon, combined with abundant evidence for lightning in volcanic columns, and anecdotal or documented observations of waterspouts, has led to a

(a)

(b)

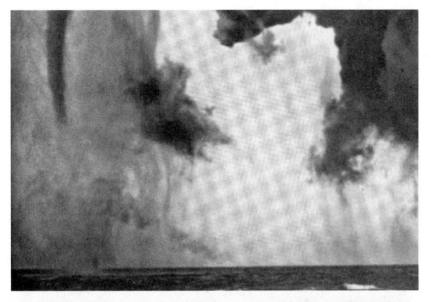

FIGURE 5.9 (a) Waterspouts spawned during the eruption of Surtsey volcano in 1963. (b) A sheath of lightning on the eruption column of Chaitén, May 3, 2008.
*Part (a): image from S. Thorarinsson and B. Vonnegut,
"Whirlwinds Produced by the Eruption of Surtsey Volcano,"* Bulletin of the American Meteorological Society
45 (1964): 440–44; *part (b) photo by Carlos Gutierrez,
Landov Media.*

theory that volcanic eruption columns have dynamics very similar to those of the mesocyclones[18] to be discussed in Chapter 8. According to this theory, the uptake of moving air from near the ground imparts a spin to the eruption column just as it does in mesocyclones, and this spin is then transferred to the umbrella, giving it the observed rotation. This interpretation opens up the possibility that atmospheric mesocyclones, which can be studied with instruments embedded in them, can provide a basis for understanding volcanic mesocyclones, in which it is nearly impossible to embed instruments.

Years after the eruption of Pinatubo ended, mudflows continued to send damaging discharges down the major drainages of the volcano. Many rivers remained clogged with sediment for years. The eruption disrupted the homes and livelihoods of more than 2 million people, destroyed 8,000 houses, and damaged 73,000 others. Total damage to public infrastructure was estimated at nearly $450 million dollars. Agriculture and forestry were seriously disrupted. Clark Air Base was abandoned by the US.

The injection of roughly 20 million tons of sulfur into the stratosphere resulted in a 10 percent reduction of sunlight reaching the Earth's surface, and a temperature decrease of about 1°F in the Northern Hemisphere for several years. Perhaps coincidentally, or perhaps not, in 1992 the US experienced its third-coldest and third-wettest summer since 1915, leading to the "Great Flood of 1993" on the Mississippi and Missouri Rivers. The injection of sulfur from the Pinatubo eruption into the atmosphere caused destruction of so much ozone that midlatitude ozone levels reached a record low, and the area of ozone depleted over the Antarctic (the "ozone hole") reached its largest size up to that time.

THE BANG IN THE BURP:
VEI, THE RICHTER SCALE OF VOLCANOES

The eruptions discussed here are relatively small compared to those that geologists know have happened in the past from their studies of volcanic deposits. Volcanologists have developed the Volcanic Explosivity Index (VEI) to provide a measure of the impact of a volcanic eruption.[19] It is based on geological knowledge that volcanic eruptions produce volumes of material ranging from trivial (a few cubic feet) to unimaginable (many thousands of cubic miles), and that they can erupt quietly or violently.

Ideally, the volume of ash produced, the height of the eruption plume above the vent, and the duration of the eruption are considered in the VEI. In practice, because prehistoric or unobserved eruptions are considered, observations of the height of many plumes, as well as measurements of the duration of eruptions, are not available, so the assignment of a VEI is based on models of these parameters. A value of 0 is assigned to small effusive lava flows (for example, a football stadium partially filled with lava would have a VEI of 0). The largest possible eruptions have been assigned a magnitude of 8—violent, colossal Hollywood fodder.

The VEI index is less quantitative than the Richter scale but nonetheless gives us a perspective on the full spectrum of possible volcanic events. It is easier to assign a VEI number to recent eruptions than to those in the distant past, but the index has been determined as well as it can be for more than 5,000 eruptions over the past 10,000 years.[20]

The 1883 eruption of Krakatoa had a plume that ascended to 15 miles and an erupted volume of more than 2.5 cubic miles of magma. Four gigantic explosions left a crater 3.5 miles across and ejected 1.2 cubic miles of magma. Land was obliterated over 5 square miles, and a tsunami over 30 feet high demolished towns and villages. More than 36,000 people died. The total energy released by the four events of the Krakatoa eruptions is estimated to be equivalent to 200 megatons of TNT (the Hiroshima atomic bomb was about 20 kilotons, or a factor of 10,000 less). Ash was so thick that the area around the volcano was plunged into darkness for days. The ash circled the Earth in thirteen days and caused spectacular red sunsets for nearly three years. During the first year after the Krakatoa eruption, global temperatures dropped by over 2°F. The eruption was assigned a VEI of 6. Pinatubo and Novarupta were also VEI 6 eruptions.

REFLECTIONS: CHAIN REACTIONS

In considering future volcanism, like other disasters considered previously, the possibility of an "unknown unknown" rears a very ugly head. Is there something nearly inconceivable out there waiting to happen? Except for Hollywood dramas of the past few decades, specifically of a possible eruption of Yellowstone,[21] the concept of an Earth-changing volcanic eruption is an unknown unknown to most people. But to geologists, this is not the case. We have reconstructed a lot about what has happened in the past, but we do not yet have all of the tools required to pinpoint when or where it may happen again, what the precursors might be, or what specific local and global consequences may result. We assume there would be warnings if Yellowstone, for example, were to awaken. But if it erupts only roughly every 500,000 years, what are those warnings and how far in advance might they occur? 1 year? 10 years? 100 years? 1,000 years? These timescales are very short compared to hundreds or thousands of years, yet potentially so long that we wonder if we will recognize a sequence of events as a precursor of something larger. This pertains not only to volcanic eruptions but equally well to earthquake forecasting.

We know that throughout recorded history, volcanoes have affected humans, their economies, and their cultures. A dormant volcano is often concealed under gentle pastoral slopes that provide rich harvests and host thriving communities ranging in size from villages to major cities. But when conditions inside the sleeping giant change, the catastrophes can be enormous, especially if the eruption sets off a chain reaction. Locally, the effects of the eruption might be restricted to blankets of suffocating ash, but might also be accompanied by a chain reaction that includes mudflows, landslides, and tsunamis (see the next chapter). Globally, the emissions can cause atmospheric temperature and chemistry changes that result in acid rain and hemispheric or global cooling.

The global changes occur largely because sulfur is emitted in volcanic plumes. Gaseous sulfur is converted to sulfuric acid in the stratosphere, returning to Earth in places as acid rain. As I noted earlier in describing the effects of the 1991 Mount Pinatubo eruption, the sulfur forms sulfate aerosols that take part in chemical reactions that contribute to destruction of the ozone layer that absorbs harmful rays from the sun. But the aerosols also reflect solar radiation, thereby helping to cool the Earth (perhaps a good, though temporary, thing in these decades of global warming). This cooling effect, however, has led to harsh conditions in the past.

The 1815 eruption of Tambora, the largest in the last 10,000 years, not only killed over 70,000 people, but contributed, by the emission of aerosols, to a global cooling of about 1°F in 1816.

That may not sound like much cooling, but it robbed North America and Europe of a summer in 1816. Snow fell in July and August in New England. One theory suggests that the effects on New England were so severe that they played a major role in the depopulation of that region because hard-hit farmers started to migrate into the Ohio Valley and midcontinent.[22] The Tambora eruption was assigned a VEI of 7.

Not all that much further back in history, the island of Thera (present-day Santorini) in the Aegean Sea had a large eruption in about 1450 BC, also a VEI 7. Initially it was believed that Thera had erupted "only" about 9 cubic miles of magma and rock, but recent estimates suggest that the volume could have been as large as 15 cubic miles. Because Thera lies fairly close to the Greek island of Crete, where the flowering Minoan civilization—sometimes referred to as the first European civilization—had developed, speculation arose that this civilization may have perished at the mercy of the Thera eruption.[23]

Going back even further, the largest eruption of the last 2.5 million years occurred at Toba, Sumatra, only about 71,000–73,500 years ago. That eruption almost certainly caused a volcanic winter.[24] Although the idea is very controversial, a number of researchers have suggested that genetic evidence suggests that this terrible time caused a bottleneck in the population with as few as 500–3,000 females surviving.[25] The Toba eruption was assigned a VEI of 8. That was truly a disaster.

Chapter 6

THE POWER
OF WATER:
TSUNAMIS

MEGA-TSUNAMIS:
A WILD RIDE IN LITUYA BAY

Late one night in 1958, when it was still light in the Land of the Midnight Sun, a landslide of ice and rocks raced down a mountain above Lituya Bay in a remote area of Alaska.[1] Slamming into the water at 250 miles per hour, it changed the state of the calm waters of the bay by generating the current world-record-holding tsunami, with a height of 1,722 feet (Figure 6.1).[2] Although it killed one couple in a boat near the shore, two other couples, in separate boats in the bay, survived to tell us about this wave.

At Lituya, glaciers had carved a deep, 7-mile-long fjord into the Alaska Panhandle during the ice ages. As the glaciers receded, they left behind a narrow pile of material (a "moraine") that formed a spit partially blocking the bay from the Gulf of Alaska. Trees covered the whole area, except the inland end of the bay, where three glaciers poured down over cliffs toward the fjord. During a magnitude 7.7 earthquake on July 9, 1958, one of these, the Lituya Glacier, disintegrated. Survivors Bill and Vivian Swan-

son, awakened by violent vibrations of their boat when the earthquake struck, recounted their experience:

> *The glacier had risen in the air and moved forward so it*
> *was in sight. It must have risen several hundred feet. I don't*
> *mean it was just hanging in the air. It seems to be solid, but it*
> *was jumping and shaking like crazy. Big chunks of ice were*
> *falling off the face of it and down into the water. That was*
> *six miles away and they still looked like big chunks. They*
> *came off the glacier like a big load of rocks spilling out of a*
> *dump truck. That went on for a little while—it's hard to tell*
> *just how long—and then suddenly the glacier dropped back*
> *out of sight and there was a big wall of water going over the*
> *point. The wave started for us right after that and I was too*
> *busy to tell what else was happening up there.*[3]

This wave was truly a mountain on the water.[4] It would have
drowned not only the Empire State Building, but anything 300
feet above it. It would have overtopped any mountain in Minnesota (1,700 feet), and dwarfed those of twelve other states and the
District of Columbia.[5]

The Swansons were moored in a small cove inside the bay, just
behind the spit. In the four minutes it took the wave to travel down
the bay to their boat, the height had decayed, but it was still big
enough to lift the boat not only up and over the spit out into the
Gulf of Alaska, but about 80 feet above trees growing on the spit.
The boat surfed the wave until the crest broke, driving the boat
to the bottom of the gulf near the spit and breaking it to pieces.

The landslide at Lituya was in some ways unusual—the result of
a dangerous combination of a glacier on a steep mountain in close
proximity to a body of water and a major active fault zone. The
plunge of 40 million cubic yards of debris—equivalent to a cube

FIGURE 6.1 Ridge bordering Lituya Bay. The top of the
ridge is at 1,720 feet, and the removal of all trees and
soil (*right*) is evidence of the height of the tsunami.
Photo by D. J. Miller, USGS.

somewhat more than three football fields in length on each side—
into the bay resembled a meteorite impact more than a typical
landslide.

More commonly, tsunami-forming landslides race down the
flanks of volcanoes, which consist of an unstable mixture of lava
flows, ash, and cinders, the latter acting like ball bearings lubricating the motion of the slide. Geological evidence tells us how
frighteningly large the tsunamis caused by such slides can be. The
tsunami from the eruption of Krakatoa in 1883 pushed the Dutch
steamship *Barouw* 2 miles up the Kuripan River, hundreds of miles
away in central Java (Figure 6.2). On the main island of Hawaii,
gravels deposited about 110,000 years ago have been found up to
200 feet above present sea level.[6] These gravels are rocks reworked
into smooth shapes by the actions of waves along the shore of the
ocean. How can they be 200 feet above sea level? A big wave?

FIGURE 6.2 The Dutch steamship *Barouw* was swept up
the Kuripan River by the Krakatoa tsunami. It came to
rest 2 miles inland, 30–60 feet above sea level. *Engrav-
ing by Edmond Cotteau, 1884. Courtesy of the Royal
Society.*

As if their presence 200 feet above the shore isn't remarkable
enough, the reconstruction of the emplacement of these gravels proves even more amazing. If Hawaii had been rising out of
the ocean over time, or if the ocean had been receding, then old
gravels could be above sea level at the present time just because
of these processes alone. But over geological time, the ongoing
volcanic eruptions in Hawaii have kept piling more and more
lava onto the surface, weighing down the island and causing
it to slowly sink. Those gravels, now 200 feet above sea level,
must have been much higher when they were deposited. Recon-
structions of the rate of sinking suggest that they may have been
deposited as much as 1,300 feet above sea level and more than
3 miles inland from the coast. Although controversial, the evidence suggests that they were dumped there by a tsunami caused

by a landslide from the flank of Mauna Loa volcano, 110,000–120,000 years ago.

A tsunami generated by a landslide, not an eruption of magma and gas, caused Japan's worst volcanic disaster, and the fifth-worst volcanic disaster in the world. During an earthquake in 1792, a prehistoric volcanic dome on Mount Unzen, on the island of Kyushu, collapsed. Sliding into the sea as a giant landslide, it drove a tsunami 330 feet high up the eastern coast of Kyushu, killing an estimated 15,000 people.

Landslides off the flanks of volcanoes have caused tragic tsunamis in Alaska, New Guinea, Indonesia, and Italy, and in the Caribbean at Montserrat. The 1980 landslide at Mount St. Helens, the largest in the US historic times, sent a big wave, sort of a local tsunami, over 800 feet high across Spirit Lake and up into the high country north of the volcano, carrying trees downed by the lateral blast into Spirit Lake as it receded.

Although catastrophic to the regions that they traverse, landslides and volcanic eruptions are not usually the processes that generate the devastating waves that travel around whole ocean basins. They are like pebbles, admittedly large pebbles, dropped into a gigantic pond. The waves that they produce simply decay too fast to survive the spreading out over long distances.

Earthquakes, on the other hand, can disturb huge regions of the ocean bottom and produce the waves that survive spreading over long distances. What determines whether a devastating tsunami will affect just a local region or hit the shores of countries across the ocean basins? To understand this, our field trip takes us under the sea around the Indian and Pacific Oceans to examine what happens to the seafloor and to the water at the site of an earthquake where a tsunami is born. We then follow the tsunami as it travels across an ocean and up onto far-distant shores by looking closely at the Tohoku tsunami of 2011 as it approached and then hit the shores of Japan.[7] The change of state relevant to this chapter is the change to the surface elevation of a body of water by an event, such as a volcanic eruption or an earthquake, and the propagation of this perturbation across the body of water. It is a change in the potential energy of water around a disturbance, and the transformation of this energy to kinetic energy of water on distant shores.

THE INDIAN OCEAN AND TOHOKU TSUNAMIS

On the morning of December 26, 2004, the magnitude 9.2 Indian Ocean earthquake sent a huge tsunami across the Indian and Pacific Oceans. Waves up to 100 feet high slammed into the coast of Sumatra,[8] and smaller, but no less deadly, waves hit the shores of countries farther away—Sri Lanka, India, Thailand, Somalia, and the Seychelles. Unfortunately, because of the remoteness of many areas, the lack of tsunami detection systems in the Indian Ocean, and poor infrastructure for communications in this part of the world, many people on these shores had no warning that their lives were about to change forever. Warnings before the tsunami struck were sparse, and eyewitness accounts from survivors are rare. This tsunami, one of the deadliest natural disasters in recent history, killed more than a quarter of a million people, possibly as many as 300,000.

In contrast, when the magnitude 9 Tohoku earthquake ripped open the floor of the Pacific Ocean in 2011, an array of sophisticated instruments on the ocean bottom detected the earthquake and tracked the progress of the tsunami that it generated as it marched toward Japan. Because Japan is located in a tectonically active area and has a long written and oral record of earthquakes and tsunamis, city managers had instituted, and enforced, strong building

TSUNAMI!

FIGURE 6.3 The nearly universal warning sign dis-
played at shorelines where tsunamis are a danger.
Image from NOAA.

codes and emergency drills. Widely distributed alert systems spread
warnings of possible tsunamis, loud sirens and loudspeakers blared
in the villages, and prominent tsunami warning signs pointed the
way to high ground (Figure 6.3). Just as earthquake awareness and
drills are common in California, tsunami awareness and drills are
common in Japan. People knew what to do.

During the Tohoku earthquake, many buildings withstood the
strong shocks and shaking of the earthquake itself. However, the
earthquake was so close to Honshu that some residents, particularly the elderly, simply couldn't flee from the tsunami to safety
fast enough. Of the 19,000 dead and missing, 65 percent were
over sixty years old. The losses from the effects of the earthquake, however, were minimal compared to the losses from the
tsunami; drowning in the tsunami claimed more than 92 percent
of the victims of the earthquake.

This region, known as Japan's "Tsunami Coast," stands out
in lists of tsunami tragedies because so many tsunamis have

repeatedly battered and destroyed it. An English observer present during a major tsunami in 1896 brought together two ancient Japanese words—*nami*, meaning "waves," and *tsu*, meaning "breaking on a harbor"—to introduce the term "tsunami," now in common use around the world. In just thirty minutes, that 1896 event, now known as the Meiji-Sanriku tsunami, obliterated 9,000 houses, carried away 10,000 fishing boats, and killed 27,000 people (Figure 6.4).

What is it like to be in a tsunami? An eyewitness to that event reported:[9]

Nearly all [of the people] were in their houses at eight o'clock, when, with a rumbling as of heavy cannonading out at sea, a roar, and the crash and crackling of timbers, they were suddenly engulfed in the swirling waters. Only a few survivors on all that length of coast saw the advancing wave, one of them telling that the water first receded some 600 yards from ghastly white sands and then the Wave stood like a black wall 80 feet in height, with phosphorescent lights gleaming along its crest . . . On the open coast the wave came and withdrew within five minutes, while in long inlets the waters boiled and surged for nearly a half hour before subsiding . . . One loyal schoolmaster carried the emperor's portrait to a place of safety before seeking out his own family. A half-demented soldier, retired since the late war and continually brooding on a possible attack by the enemy, became convinced that the first cannonading sound was from a hostile fleet, and, seizing his sword, ran down to the beach to meet the foe.

Anyone who has played in the 5-, 6-, or 10-foot waves of the ocean surf has almost certainly been slammed to the ocean bottom

FIGURE 6.4 The 1896 Meiji-Sanriku tsunami inundated a mountain pass that now bears this elevation marker (38.2 meters is 125 feet). *USGS photo by Bruce Richmond.*

by the force of those waves. The very rare, highly skilled, danger-loving surfer can ride a 100-foot wave under perfect surfing conditions, but any miscalculation results in tragedy. No one could have ridden these tsunamis. One wave in the Meiji-Sanriku tsunami held the world record at 125 feet until it was eclipsed by the 127-foot-high wave recorded during the Tohoku earthquake. How big are these waves? They are about the height of the Watergate complex in Washington, DC, or the Christ the Redeemer statue that looms over Rio de Janeiro, Brazil. Because of the long history of tsunamis (and severe typhoons) in Japan, seawalls, some as high as 30 feet, protect many coastal villages. Unfortunately, these were terribly inadequate in 2011, when they were simply overwhelmed as massive amounts of water marched inexorably higher and higher over the walls and through the cities that they failed to defend.

The life of a tsunami consists of three stages: birth, propagation, and death. A tsunami is born during a particularly dangerous change of state—the disruption of the normal state of the ocean by an earthquake, landslide, or volcanic eruption. It propagates across the oceans, amazingly long distances, before dying on distant shores.

What causes tsunamis? Does every earthquake generate a tsunami? Why are they so devastating? How are the waves of tsunamis different from ordinary waves on the beach? Why do they affect places so far away from where they originate? Before we can answer these questions, we need a brief diversion into the general properties of waves, a diversion that is relevant not only here, but also in the chapters that follow.

PRIMER: WAVES AND TEENAGERS

What is a wave? Waves carry information about a change somewhere as they travel through a material. The change is called the "generating force," and there are many different types of generating forces. Just as an earthquake sends seismic waves through the earth (Chapter 3), a whisper from lips sends sound waves through air, the impact of a stone on a pond sends surface waves running out across water, a bolt of lightning on a dark and stormy night sends shock waves out through the atmosphere, and a traffic accident sends collision waves through cars on a freeway. Many scientists and engineers, myself included, spend their entire career trying to figure out what waves can tell us about generating forces.

When a generating force has created a wave, other forces work to oppose it and cause it to die out. These opposing forces include friction, gravity, surface tension,[10] and viscosity. When, for example, the impact of a pebble on a pond generates a crater in the water, gravity works to fill the crater and to restore the water in the pond to its original undisturbed state. Waves in which gravity is the restoring force are, appropriately, called "gravity waves."[11] Tsunamis are gravity waves, and they also belong to a class of waves called "traveling waves." These typically occur when a wave is free to travel long distances without much interference, as is the case, for example, when a tsunami spreads out across the ocean. (Standing, or "stationary," waves, by contrast, tend to form when a wave is trapped in a confined space, like a bathtub in contrast to an ocean.)

Waves are rather like teenagers, in that they usually (but not always) travel in groups. These groups, or "sets," of waves have characteristic properties: height, steepness, length, frequency, and period—the latter three of which are interrelated (Figure 6.5). And like teenagers, water waves come in all sizes and shapes, but the British mathematician-physicist George Gabriel Stokes showed us a way to simplify their analysis. In the mid-1800s, the science of fluid dynamics was in its infancy. Then, at the age of twenty-three, Stokes published key papers that were to dramatically improve our understanding of fluid motions in general, and of waves in particular. After a lot of approximations and elegant mathematics, Stokes realized that by comparing the water depth to the length of waves, two very different behaviors could be recognized.[12] Water depth is described as either "deep" or "shallow," but these terms have meaning only in the context of a specific wave.

If the depth of water through which the wave is traveling is greater than half the wavelength, it is considered "deep," and waves that meet this criterion are called "deep-water waves" (which we'll talk about in the next chapter). If the depth is less than one-twentieth of the wavelength, the water is considered "shallow," and waves that satisfy this criterion are called "shallow-

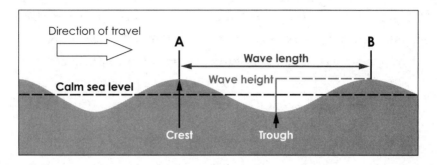

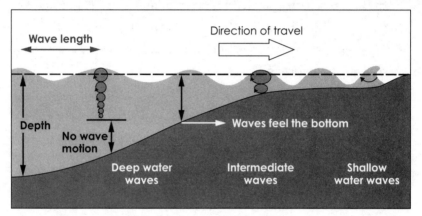

FIGURE 6.5 Schematic illustrating the terminology used to describe waves.

water waves." Note that, in this context, (1) water that we consider to be "very deep," such as in the oceans, might be "very shallow" in a fluid-dynamical sense if wavelengths are sufficiently big, and (2), as we explore in this chapter, tsunamis behave as shallow-water waves as they traverse even the deepest oceans. In an ocean that is 2 miles deep, for example, waves with lengths of 4 miles or less are deep-water waves, and waves with lengths greater than about 80 miles are shallow-water waves. There is quite a range of wavelengths between the deep- and shallow-water limits— between 4 and 80 miles in this example. Waves with lengths in this range are difficult to analyze because they don't satisfy either the deep- or shallow-water approximations, and we will not consider such waves any further in this book.

With this primer on waves under our belts, we can return to examining tsunamis.

BIRTH OF A TSUNAMI

When there is a tsunami alert, officials send two messages to the public: go to high ground, and stay there even after the first wave has passed until we give the all clear. Why? A tsunami is often not just one wave, but can consist of a series of waves with wavelengths of 100 miles or more. Because this wavelength is so long, successive waves—which might be higher than the first wave—may arrive minutes or even many tens of minutes after the first wave. To understand why tsunamis have this character, we have to look at changes of state that occur over just a few seconds or minutes in the Earth's crust and its overlying oceans at the source region of the earthquake.

Not all earthquakes generate tsunamis, and the uncertainty causes nightmares for tsunami forecasters. When should a warning be issued? When is a warning not warranted? To forecast whether or not a tsunami will form and how big it might be, geophysicists and oceanographers work together with complicated data analysis and modeling software on huge computers that continuously monitor seismic signals coming through the earth.[13] It takes about twenty-five minutes for seismic waves from an earthquake to reach all stations of the global seismic arrays, and another ten minutes for computers to crunch the data through models. Officials can now issue a preliminary estimate of tsunami danger in less than an hour after a quake.

During this hour, scientists are searching for four pieces of

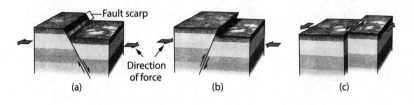

FIGURE 6.6 Schematic illustrating the terminology used to describe fault motions that are tsunamigenic (a,b) and those that are not (c).

information that determine the likelihood of a tsunami: the geometry of the fault motion, the energy and depth of the earthquake, and the depth of water over the fault rupture zone. Let's examine briefly why these four quantities matter.

The last one is easy. The amount of water entering a tsunami is determined by the water depth and the area of the crust displaced during an earthquake. The more water above the displaced crust, the more water that moves out into the tsunami. The area of displaced crust is in turn related, in a rather complicated way, to the geometry of the fault motion.

Tsunamis form during quakes in which rocks have some vertical motion along a fault, causing a bulge or depression to form in the water (Figure 6.6). Quakes with a large component of vertical motion (see Figure 6.6*a,b*), known as "tsunamigenic" earthquakes, occur in the subduction zones of the world, where one plate of the lithosphere is descending under another (see Figure 3.1). Tsunamigenic earthquakes occur primarily around the Pacific Rim, where two-thirds of all tsunamis occur, and also in the Indian Ocean, the Caribbean, and the Mediterranean. Earthquakes in which there is little vertical motion—those that have predominantly "strike-slip" motion (see Figure 6.6*c*)—do not generate tsunamis. The San Andreas Fault in California is an example of this type of fault.

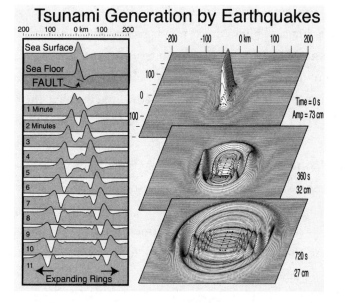

FIGURE 6.7 Computer simulation of the behavior of the surface of water over a magnitude 7.5 earthquake caused by motion along a thrust fault. The surface expression of the fault zone is indicated by the dashed rectangular box. The initial water displacement is shown at time = 0, and the waves are shown at six and twelve minutes. Note that the biggest waves and most energy are propagated perpendicular to the fault plane where it intersects Earth's surface. *Courtesy of Steve Ward, modified from S. N. Ward, "Tsunami," in* Encyclopedia of Solid Earth Geophysics, *ed. H. K. Gupta (New York: Springer, 2011), 1493–98.*

When a fault tears the crust of the earth during a tsunamigenic earthquake, a piece of the crust, thousands of square miles in area, suddenly rises up or drops down. During the Tohoku earthquake, an area of nearly 6,000 square miles—the size of Connecticut—rose up as much as 15 feet, disturbing an enormous volume of water.[14] If the crust rises, it pushes a bulge of water upward, and this bulge of water responds to the force of gravity by flowing out into the surrounding ocean. If the crust drops, the overlying water also drops, forming a depression in the water surface. The ocean responds to gravity by flowing into the depression. Such events give birth to a tsunami (see Figure 6.7).

To understand how the long wavelengths and multiple waves of a tsunami develop during its birth, imagine that the rupture of the crust during an earthquake creates a piston made out of a piece of the crust. A typical piston might be a hundred miles long and tens of miles wide. For the sake of discussion, assume that it thrusts upward during an earthquake. Most of the energy released during the quake goes into crushing and heating rocks adjacent to the fault, but a small fraction of the energy is transferred to the water. This energy can, however, be enough to set a huge volume of water into motion.

The process of transmitting the energy to the water is both related to the area of the piston and how far it moves during the quake—its displacement. (Boxers understand this piston physics well—a big fist thrown in a long punch is more powerful than a small fist with a short reach.) Earthquake magnitude is also related to the area and displacement of the crust, and this is the reason that estimates of the magnitude of the earthquake are so important in predicting the size of a tsunami.[15]

The potential for devastation by a tsunami depends on the fault geometry in one other way. Waves from a fault rupture are bigger in some directions than in others. Waves spreading away from the long sides of a rupture—that is, spreading perpendicular to the fault plane—tend to be much bigger than those that spread away from the tips of the fault, parallel to the fault plane (see Figure 6.7). If two coastlines are at equal distances from a fault, the one perpendicular to it is more likely to experience damage than the one parallel to it. Because of this geometric effect, Bangladesh— though lying nearly at sea level—was spared much or the possible devastation from the 2004 Sumatra earthquake.

The depth of an earthquake matters because a tsunami is produced only when the seafloor moves into the overlying ocean. If the earthquake is buried too deep in the crust, the fault that caused it may never break through to the seafloor or, if it does, may be moving too sluggishly to give the water a good punch. Shallow earthquakes with substantial vertical displacements and high moment magnitudes generate the largest tsunamis.

According to at least one model,[16] the dynamics of the first few seconds after an earthquake are dramatic. In fact, they are so dramatic that I suspect it would be no fun at all to be a fish near the source of a tsunamigenic earthquake! The acceleration of the crustal piston during the rupture creates a pressure pulse in the overlying water—in much the same way that a piston in the cylinder of an automobile creates a pressure pulse in the gas-air mixture of a combustion engine.

The initial pressure pulse spreads, at a velocity of nearly a mile per second in all directions through the ocean. In horizontal directions, it heads toward far or near coastal shores. In the vertical direction, however, the pressure pulse heads up toward the surface of the water that is, at most, a couple of miles away. When this pressure wave hits the water surface, it reflects back toward the ocean bottom, where it, in turn, is reflected again toward the surface—bouncing back and forth in much the same way that optical or acoustic waves reflect off of obstacles in their paths. Such reflections restore the ocean to a condition of equilibrium with the new configuration of the seafloor produced by the earthquake.

A specific example, greatly simplified so that a supercomputer could crunch through the equations, provides a glimpse into this process.[17] This example is a miniaturized version of the real process: the ocean is only a half mile deep and the crustal piston is only about 100 yards in diameter—simplifications that help reduce the required computer time. The pressure of the water on the ocean bottom is about 100 times normal atmospheric pressure. The piston is given an initial shove upward by the earthquake; in only five-thousandths of a second it reaches a vertical speed of about 450 miles per hour. (For comparison, it takes a top fuel dragster much longer, seven-tenths [0.7] of a second, to accelerate from the starting gate to a mere 100 miles per hour.)

The punch of this piston increases the pressure in the water at the seafloor by about 4,000 times atmospheric pressure—a factor of 40 over the normal pressure exerted by the weight of the water. No fish in the vicinity would like that much of a pressure increase, which is much more than it takes to rupture a human eardrum. Speeding out in all directions at nearly a mile per second, it takes the wave a bit over a half second to reach the surface directly over the piston. At the water surface, the wave reflects back down toward the bottom of the ocean, setting up the series of reflected waves that eventually restore the normal pressure distribution in the ocean. In this example, restoration of normal pressure distribution takes about sixteen seconds.

When the pressure wave hits the water surface above the piston, it forms a bulge—the tsunami—which in this particular example is 8 feet high. In the sixteen seconds of disturbance in this hypothetical case, waves spreading out into the ocean expand the diameter of this bulge to about 30 miles. The process that produces the bulge sets in the wavelength of the tsunami—30 miles in this example, but much larger in real earthquakes with a deeper ocean above them. The waves resemble those illustrated in Figure 6.7.

We know, however, that earthquakes that produce huge tsunamis do not have instantaneous ruptures; the ruptures can last for minutes. The waves that spread out from the source are actually very complicated. One way to extend the simple model described here to such a prolonged rupture is to think of it as consisting of a series of small, simple, instantaneous events produced at different sites along the fault. From this way of looking at the problem, we can see that all of the waves produced will have long wavelengths because the source regions, a series of the hypothetical pistons described earlier, are so large.

The simple example illustrates three critical points about the birth of a tsunami. The first is that the size of the zone of disturbed water, which establishes the wavelength of the tsunami, is determined by detailed events lasting seconds to minutes at the source. The second is that that wavelength—tens to hundreds of miles—is very large compared to a typical ocean depth (a few miles). The third is that all of the water over the disturbed crust, from the bottom of the ocean to the surface, is set into motion— not just a superficial layer on the top or bottom of the ocean. These three characteristics have important consequences for the travel of the tsunami from its birthplace to distant shores.

RUNNING FREE: TSUNAMIS AT SEA

As we've already seen, the tsunami birth process generates waves with very long wavelengths, much longer than the depth of the ocean. Tsunamis therefore travel as shallow-water waves, according to our wave primer. The velocity of such waves depends only on the water depth (more accurately, on the square root of the product of the water depth multiplied by the value of the gravitational acceleration for Earth). The Tohoku tsunami roared east across parts of the Pacific Ocean that have an average depth of about 2.5 miles. Its average velocity was about 450 miles per hour,

the speed of a commercial jet. The part of the tsunami propagat-
ing to the west toward Japan crossed through shallower water,
but it still raced toward Honshu at a rate of about 250 miles per
hour. Honshu was only 125 miles away, and although this may
sound like a long distance from an earthquake, it's frighteningly
short when it's the distance between people living on a coastline
and the site of a tsunami-generating quake. It took less than a
half hour for the tsunami to reach the coast.

A characteristic of shallow-water waves is that all of the water
surrounding the source, from the bottom to the top of the ocean,
is set into motion. As the crest of a wave passes, particles move
forward in the direction of the wave, and as the trough passes,
they move back in the opposite direction (see Figure 6.5). On average, after the wave has passed, the particles are about where they
were before the wave arrived. But because tsunamis have such long
wavelengths, the water particles can travel a long distance one way
or the other before returning to their original positions. When, for
example, the crest of a tsunami moves onshore, it can propagate
a mile or more inland before the trough behind it catches up and
starts to move the water back out toward the ocean.

Ironically, tsunamis are rarely observable at sea. Except in the
immediate vicinity of the epicenter of a tsunamigenic quake, the
height of even the largest tsunamis at sea is very small compared
to the wavelength. This means, for example, that any ship riding
over such a wave would have no sensation that the wave had even
passed. Even the great Tohoku tsunami was less than a foot high
as it traveled along most of its path across the Pacific toward the
coast of North America.[18]

A tsunami travels through parts of the oceans where water
depth varies, encountering islands and continents along its path.
It bends around and reflects off of obstacles, much like musical
waves from a symphony orchestra reflect off of the architectural

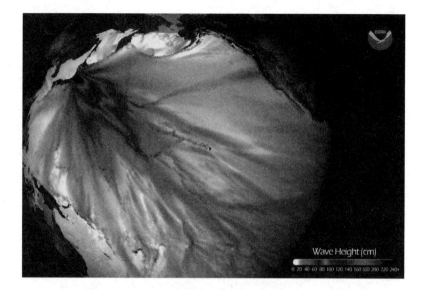

FIGURE 6.8 NOAA model of the tsunami wave height
as it propagated eastward through the Pacific Ocean
basin. The darkest colors correspond to waves 10–20
feet high, but over most of the Pacific, the waves were
less than a foot high (lighter colors). The area of a model
such as that depicted in Figure 6.7 would be invisibly
small at the source region in the scale of this figure.

elements in a well-designed music hall. These encounters affect
both the speed of the tsunami and its height. One of the most
striking visualizations of this phenomenon was the predicted
wave height of the Tohoku tsunami calculated by scientists in the
National Oceanic and Atmospheric Administration (NOAA).
Looking like an eyeball during a hangover (Figure 6.8), the differ-
ing shades of gray reflect the heights of the waves as the tsunami
reached different positions in the Pacific Ocean.

The massive destructiveness of a tsunami is directly related to
the huge volume of water moved by it. A typical tsunami contains
more than 50,000 times the water in even the biggest wind-driven
waves.[19] The power in a tsunami is likewise immense. In fact, the

power, in a strict scientific sense, can be calculated. The 2004 Indian Ocean tsunami had a power of about a million watts, a megawatt, per yard of shoreline.[20] This translates to nearly 2 (1.7) billion watts per mile of shoreline. Nearly 300 miles of shoreline were heavily impacted, which would make the total power of the tsunami 550 billion watts. Since there are about 7 billion people on the planet, this power could have lit a 60-watt lightbulb for every person on the planet for the duration of the tsunami.[21]

DEATH OF A TSUNAMI: RUN-UP

The tragic 2004 Sumatra earthquake and tsunami emphasized the need for systems in which seismic data about the earthquake, models of tsunami generation and propagation, and new instrumentation could be combined into a hazard model and real-time warning system. Such instrumentation was in place around Japan in 2011. In particular, two ocean-bottom pressure sensors provided critical data about the Tohoku tsunami.[22] The sensors were, respectively, 47 and 34 miles from the coast near Kamaishi, one of the cities most heavily devastated by the tsunami. The records give an extraordinarily detailed picture of the height of the tsunami as it passed over the sensors. During the first thirteen minutes, the height rose to 7 feet above normal sea level. Then, instead of abating, even more water piled on—an additional 10 feet over the next two minutes (100 seconds to be exact), giving a total height of 16–17 feet in these near-coastal locations. Only then did the elevated sea level drop back toward "normal" over several more minutes.

Closer to shore, we know that the tsunami was even higher. Susumu Sugawara, an experienced abalone fisherman, heard of the approaching tsunami. Instead of running inland, he ran to his boat, freed it, headed out to sea, and rode over the incoming

tsunami. His report gives us one of the very few close-up eyewitness accounts of a tsunami in the shallow waters just a few miles
from shore:

> *My feeling at this moment is indescribable . . . I talked to*
> *my boat and said you've been with me 42 years. If we live or*
> *die, then we'll be together, then I pushed on full throttle . . . I*
> *climbed the wave like a mountain. When I thought I had got*
> *to the top, the wave got even bigger.*[23]

He thought that there were four or five waves and that they were
over 60 feet high.[24]

How did these waves, so negligible on the deep ocean, become
100-foot monsters as they approached the shore? It all comes
down to those conservation laws of mass, momentum, and energy
that we reviewed in Chapter 2.

Waves traveling from the deep ocean toward a coast gradually
encounter the shallower waters of the continental shelf and coastline. Since the speed of the waves depends on water depth, they
slow down as the water becomes shallower. In fact, they deceler-
ate from the speed of a jet aircraft to nearly walking speed over a
fairly short distance (depending on the detailed geometry of the
sea bottom). Since the waves slow down, the wavelength shortens,
and consequently, from conservation of mass and momentum,
the waves grow in height, becoming the destructive monsters that
devastate coastal communities (Figure 6.9).

One of the most surprising things about tsunamis is how long
they keep building up. A tsunami rarely forms the spectacular wall
of water that is depicted in some movies. Rather, tsunamis are much
more like cycles of high and low tides, all occurring within tens of
minutes instead of twelve hours like natural tides do. From a Dutch
pilot stationed at Anjer on Java came this report after Krakatoa:[25]

FIGURE 6.9 Growth of the height of a tsunami as it enters shallow water near shore. The height increases because the velocity of the wave depends on water depth, and the tsunami slows down as the water becomes shallower toward the shore. *Adapted from http:// web.mit.edu/12.000/www/ m2009/teams/5/research.html.*

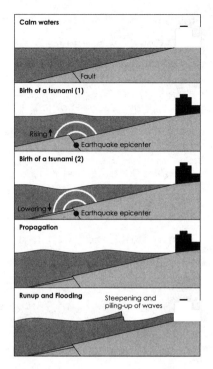

Calm waters

Fault

Birth of a tsunami (1)

Rising↑
● Earthquake epicenter

Birth of a tsunami (2)

Lowering↓
● Earthquake epicenter

Propagation

Runup and Flooding   Steepening and piling-up of waves

The huge wave rolled on, gradually decreasing in height and strength until the mountain slopes at the back of Anjer were reached, and then, its fury spent, the waters gradually receded and flowed back into the sea.

As the 2011 Tohoku tsunami ran onshore and interacted with urban infrastructures, its character changed dramatically. The white-water foaming waves at sea were transformed into dense, turgid, sluggish debris flows laden with materials scoured along their path. Such flows had more in common with watery, muddy landslides and debris flows (discussed in Chapter 4) than with the surfing waves that we associate with the normal conditions along our coastlines. The transformation was particularly vivid where the tsunami came onshore at Sendai, a town of a million people. Approaching the shore, the tsunami gouged up mud from the bottom of the harbor and then, coming onto land, marched inexorably down the runways of the Sendai airport, engulfing helicopters and planes, both small and large, eventually crushing them. Over a distance of less than a mile, the watery tsunami changed into a massive, dark, turbid flow of debris and took on a different character along each side of the airport defined by the details of what it encountered along its path.

Tsunamis do not attack only coastlines. They work their way inland along any low parts—in particular, along the estuaries and rivers that normally flow toward the ocean. During the Tohoku tsunami, in some cities river water flowing toward the ocean was overridden by the tsunami flowing the opposite direction. It picked up ships in port and tossed them around into each other in a slow-motion ballet (Figure 6.10). It moved them chaotically inland, sometimes under bridges that still carried cars and people moving over the churning waters below. Thousands of cars from parking lots, manufacturing plants, and shipping docks were swept together, creating an eerie scene that looked as if the normal driftwood on the ocean had been replaced by cars and pieces of cities. The auto company Nissan alone reported 2,300 cars destroyed by the tsunami. Mats of cars, floating buildings, and debris from crushed buildings were carried in a slow march across highways and farm fields. Burning buildings floated on the mat as if unaware of the watery conveyor belt below. An oil refinery or chemical plant exploded into a fireball. As the tsunami encountered coastal cities, it split into individual streams of roiling water funneled down the urban corridors, destroying the infrastructure and the lives of tens of thousands of victims. In all, more than 27,000 people died, and more than a million homes were destroyed. However, without the planning, practice, and discipline of the citizens and their leaders, the toll would have been

FIGURE 6.10 A ferry rests on top of a two-story building in Otsuchi, Iwate Prefecture. *Reuters / Toru Hanai
image RTR2LBJ9.*

much greater. Their actions before and during this horrible event
and their incredible recovery are a model for the rest of the world.

REFLECTIONS:
WHERE, WHEN, BUT NOT "IF"

Although mega-tsunamis can be generated by landslides and
volcanic eruptions, those generated by earthquakes are our most
immediate threats. It's just a matter of time and place. Geoscientists bring a wide array of tools to the task of predicting time
and place—knowledge of geologic setting and seismic patterns,
measurements of crustal deformation, and analysis of historic
tsunamis and their deposits. Two examples of this approach illustrate our current state of understanding—one from an analysis
published in 2001 that warned of a high probability of an event

like the 2011 Tohoku earthquake and tsunami, and another published in 2007 that predicts an earthquake and tsunami that we
still await off the coast of Myanmar (formerly Burma).

On July 13, AD 869, a date known so accurately because the
historical record in Japan is so well preserved, a tsunami struck
the Sendai region just described in the last section. Historic
records like this are of tremendous help to geologists as they try
to unravel sand deposits caused by storms and their rogue waves
(see the next chapter) from those due to tsunamis. This tsunami,
now called the "Jōgan tsunami" after the reigning emperor at the
time, flooded an enormous area and carried sand from the coast
as far as 2.5 miles inland. A reconstruction of the conditions indicates that an earthquake of approximate magnitude 8.3 caused
the tsunami. By working through the deposits and geologic setting, scientists inferred a location for the earthquake, estimated
its size, and constructed a model of the fault motion during the
quake.[26] A model of the tsunami that would have been generated
suggested that waves with heights at least 25 feet were the cause
of the extreme inland inundation and extensive sand deposits.
They further noted that two similar deposits could be identified
in this area—one with a date sometime between 140 BC and AD
150, and one during the interval 910–670 BC. The three tsunamis
are separated by roughly 800–1,100 years, an observation that
led the scientists to state (in 2001): "More than 1100 years have
passed since the Jogan tsunami and, given the reoccurrence interval, the possibility of a large tsunami striking the Sendai plain is
high. Our numerical findings indicate that a tsunami similar to
the Jogan one would inundate the present coastal plain for about
2.5 to 3 km [1.6–1.9 miles] inland."[27] Ten years later the Tohoku
tsunami struck.

The great Sumatra-Andaman earthquake and tsunami of 2004
also had at least one historical relative, the April 2, 1762, Arakan

earthquake that occurred in the northern Bay of Bengal along the coast of Myanmar.[28] This area is flanked on the west by the part of India near Calcutta, and on the north and northeast by Bangladesh and Myanmar. It is a geologically complex place, but the tectonic setting is similar to other subduction zones that have generated tsunamigenic earthquakes. Perhaps most importantly, there are observations of stress and strain in the crust that indicate that the faults here are locked and accumulating stress.

Captain E. Halsted, on a British survey expedition in this area in 1841, recorded clear evidence that there had been 10 to 23 feet of recent uplift along the coasts near Myanmar. Stories relayed by locals told that their fishing practices had changed due to the uplift, all presumed to have occurred in 1762. Computer simulations using the reconstructed earthquake scenario show that a large tsunami in the future will again tear through this region. Chittagong, the second-largest city in Bengladesh with a population of 6 million, lies right over the fault. The Ganges-Brahmaputra delta is home to over 60 million people living within 33 feet of sea level. If a few percent of the population is vulnerable to a tsunamigenic quake and tsunami, the author of the analysis projects that over a million lives are at risk. Large earthquakes in this region tend to recur about every 500 years. With 250 years already elapsed since the 1762 earthquake, it may well be another 200–250 years before a repeat of the 1762 earthquake, but it also may not be that long before a somewhat smaller, but still enormously lethal, one does strike.

The positive side of these case studies is that they illustrate that we know a lot about earthquakes and tsunamis, and that we know how to study them. On the other hand, forecasts such as these that have large windows of uncertainty are limited in their direct usefulness. We discussed a similar problem in the

Reflections section of Chapter 3 with the use of statistics and probability.

These tsunamis cannot be prevented, but surely the deaths of hundreds of thousands or a million people can—even given the large uncertainties of predicted place and time in forecasts. Development and deployment of monitoring systems to detect tsunamis as they are born and to follow their progress across oceans, combined with speedy warning systems, public awareness, wise decisions about land use, appropriate construction practices, and disaster drills, can all greatly reduce future loss of life in such regions.

Chapter 7

ROGUE WAVES, STORMY WEATHER

OOPS . . . MY HELICOPTER IS TOO LOW!

Tsunamis, although disastrously big onshore, are tiny on the open ocean compared to some monster "rogue waves" that roam the oceans (Figure 7.1). Nowhere is this better known than at the site of the "toughest yacht race in the world," which takes place offshore of Australia. On Sunday, December 27, 1998, during the famous Sydney–Hobart maxi-yacht race, six people died, five boats had to be abandoned, and two sank during a ten-hour storm. Of the 115 boats that started the race in Sydney, only 44 made it to Hobart. At one point in the rescue effort—the largest sea rescue operation in Australian history—the pilot of a helicopter hovering 100 feet above the sea surface saw an enormous approaching wave. Climbing as fast as possible to 150 feet, he noted that the altimeter showed that he had cleared the wave by 10 feet. The wave was 140 feet high,[1] the largest rogue wave for which an actual measurement exists.

Like tsunamis, rogue waves are a manifestation of a change of state in the ocean surface, but in their dynamics these waves

FIGURE 7.1 *Great Wave off Kanagawa*, by Katsushika Hokusai. Usually interpreted as a tsunami, the wave in this print is actually a rogue wave off Kanagawa Prefecture south of Tokyo, Japan, an area of notoriously dangerous shipping conditions in the Kuroshio Current.

are completely different from tsunamis. To investigate how rogue waves differ from tsunamis, our field trip takes us to a number of places in the midlatitudes, 30°–60° north and south of the equator, to look at the dynamics of the atmosphere-ocean system and why rogue waves can form where they do.

Photographic documentation, such as that in Figure 7.2, and scientific data on individual rogue waves are still sparse, but physical evidence of their power does exist (for example, Figure 7.3). In 1861, a wave broke glass windows 85 feet above the ground in an English lighthouse—after climbing up a 130-foot-high cliff! This would imply that the wave was 215 feet high. As of the publication of this book, no wave near this height has been documented by eyewitnesses or with instruments.

FIGURE 7.2 A giant wave in the Bering Sea pounds the NOAA ship *Discoverer* in 1979. *Photo by Commander Richard Behn, NOAA.*

FIGURE 7.3 A monster wave engulfs a lighthouse, location unknown. *From NOAA's National Weather Service Collection, Mariners Weather Log.*

Rogue waves are also known as "freak waves," "monster waves," or "mad dog waves."[2] They can occur in the form of a single isolated gigantic wave, or as a set of several big waves. For much of mariner history, there were questions about the existence of rogue waves and, especially, about their size. As of 2009, the number of rogue waves recorded was only in the hundreds,[3] and I can testify after doing the research for this chapter that there are only a few good photographs of them. Only in recent decades have instruments on buoys and drilling platforms provided firm evidence that these waves do exist, that they are indeed huge, and that, while rarely observed because of the immensity of the oceans, they are common enough to pose significant hazards.

Rogue waves also occur on some large inland bodies of water, such as the Great Lakes.[4] The weather, including violent wind, that storms around the Great Lakes in the autumn is known as the "Witch of November." The storms can be as intense as category 1 and 2 hurricanes. In October 2010, a storm blasted Duluth, Minnesota, with 81-mile-per-hour (mph) wind gusts and 19-foot-high waves. In 1913, a November storm, often referred to as the "White Hurricane," battered the Great Lakes, overturning ships on four out of the five Lakes, killing more than 250 people, destroying 19 ships, and costing about $100 million (in 2010 dollars).

On a bitterly cold night in November 1975, the SS *Edmund Fitzgerald*, a freighter carrying a full cargo of ore across Lake Superior, suddenly lost all contact with the world and sank to the bottom. At the time, a massive winter storm was churning up the lake, with winds of 60 mph, gusts to 90 mph, and waves to 35 feet. The captain of a second ship, the SS *Arthur M. Anderson* in the vicinity of the "Mighty Fitz," gives us a clue about the probable cause of her sinking. Of his own ship, he reported two monstrous waves. The first hit the ship on the stern and worked its way forward along the deck, driving the bow down into the sea. Then, reported the captain, "the *Anderson* just raised up and shook herself off of all that water—barrooff—just like a big dog. Another wave just like the first one or bigger hit us again. I watched those two waves head down the lake towards the *Fitzgerald*, and I think those were the two that sent him under." Taking twenty-nine crew members with her to the bottom, the Mighty Fitz remains the largest ship ever lost on the Great Lakes. In popular culture, the event was captured by Gordon Lightfoot in "The Wreck of the *Edmund Fitzgerald*," his second most popular song and one that reached number one status in popularity in Canada and number two in the US.

One relatively well-documented rogue wave was encountered on the North Atlantic during World War II by the RMS *Queen Mary* ferrying 16,082 American troops across the Atlantic to Europe (a record that still stands for the most passengers on a ship). During a gale 700 miles off the coast of Scotland, a rogue wave 90 feet high slammed into her, tilting her to an angle of about 52 degrees (the "list" of the ship), only 3 degrees short of the 55-degree list that would have caused her to capsize.

Although the general public rarely hears reports of big waves sinking ships, accidents involving rogue waves do make the news a few times a year. Headlines in 2010 told of two men killed when a set of three "abnormal waves," all greater than 26 feet high,[5] struck a Greece-based cruise liner, the *Louis Majesty*, carrying 2,000 passengers off the coast of Spain. That same year, in a highly publicized event, sixteen-year-old Abby Sunderland set sail from Cabo San Lucas in an attempt to become the youngest person to sail solo around the world. Four months and 12,000 nautical miles after she started out, her dream ended in the

Indian Ocean, about 2,000 miles west of Australia, when "one long wave" hit her boat, rolled it 360 degrees, destroyed its mast, and left her without any means of control.

Whereas the reasons for monitoring and modeling tsunamis are obvious to the general public, the reasons for doing the same for rogue waves are not so clear. Tsunamis directly and visibly impact people and the infrastructure of their communities, but rogue waves are out of sight and mind to nearly all but mariners and ocean tourists. However, they pose a threat to three very expensive industries: shipping, fishing, and oil and gas production. Shipping includes the many oil supertankers that travel from the oil-rich Persian Gulf states around South Africa, one of the more dangerous areas. The strongest ships are designed for waves 33–50 feet high, well below the size of many rogue waves. Oil and gas production takes place on platforms in the stormy North Atlantic and North Sea.

Lloyd's Register, based in London, has a database of ship accidents and their location from which researchers can find information about ships from 1890 to the present. In 1910, one ship in every 100 was lost, but by 2010, improvements in ship construction and navigation equipment had reduced this number to one in every 670. Most accidents occur fairly close to coastlines, although there are a significant number in the open Atlantic Ocean between latitudes 0° and 60°, partially reflecting the predominance of shipping routes in these areas. During the eleven-year period from 1992 to 2003, 1,049 merchant vessels were lost from various causes out of a total world fleet of nearly 40,000—an amazingly high average of nearly 100 vessels per year.[6] A bulk carrier or tanker was lost roughly every other week. During the period of the study, foundering (direct sinking, in contrast to war, equipment failure, and so on) was the major cause, accounting for 30.9 percent of incidents. Weather and waves are the presumed culprits.

What distinguishes a rogue wave from a merely "big" wave? What forces whip a body of water into such a frenzy that rogue waves arise? In their dynamics, how do these waves differ from tsunamis? Where do they occur, and why? Before addressing these questions, we need to know a bit about how the dynamics of water in wind-driven waves differ from the dynamics of tsunamis that we discussed in the last chapter.

AN ASIDE ABOUT WIND-DRIVEN WAVES

Wind-driven waves are the "deep-water waves" George Stokes mentioned in the previous chapter and illustrated in Figure 6.5. In these waves the water depth is much greater than half the wavelength. What effect do such waves have on the water through which they pass?

I figured this out when, in the 1950s, it was a big deal to take the annual vacation from landlocked northwestern Pennsylvania to the New Jersey shore. Swimming out into the ocean, I would turn around and face the shore, treading water as the waves took me up and down, toward and away from the beach. I imagined that I had pencils sticking out of my ears, and discovered that the imaginary pencils traced a circle—the top of the circle when a crest came and lifted me up and toward the shore, and the bottom when the next trough came past and pulled me down and back away from the shore. The diameter of the circle was about the height of the waves, trough to crest. In hindsight, I had discovered one of the fundamental characteristics of particle motion in wind-driven waves: the waves passed by and broke on shore, but I, and the water in the waves, just circled in orbit around a stationary point (see Figure 6.5).

Like surfers and others playing in the waves along the shore, I also discovered that the motion of water driven by these waves decreases dramatically with depth (strictly, it decreases exponentially with depth). If an ominously big wave came toward us, or if we were caught in the surf zone of breaking waves, we knew that if we simply dove down a few feet into quieter water under the surface and let the wave pass over us, we would be safe. We were playing with waves typical of a normal shoreline. Their spacing (wavelength) was a few tens to, at most, a few hundred feet. The wavelengths of wind-driven waves on the ocean are rarely greater than 500 feet, and thus, at depths of a few hundred feet, water is calm even when waves in wild and violent seas surge back and forth at the surface.

The dynamics of wind-driven waves are fundamentally different from those of tsunamis. Whereas tsunamis, as shallow-water waves, travel with a speed that depends on ocean depth, deep-water waves travel with a velocity that depends, instead, on their wavelength (or period). One way to understand this is to recognize that since deep-water waves do not affect ocean waters below about half their wavelength, they are unaffected by the ocean bottom, no matter how deep it is. Their velocity, therefore, cannot depend on the water depth.

Deep-water waves that have a long wavelength travel faster than those that have shorter wavelengths. This dependence of wave speed on wavelength has enormous implications for the ocean response to storm systems. Storms at sea generate waves with many different wavelengths. Those with the longest wavelengths travel fastest, so they move away from the storm area faster than do waves with shorter wavelengths. Waves of different wavelengths separate along their journey across the ocean—a phenomenon known as "dispersion." Surfers know this phenomenon very well:

swells with long wavelengths, perfect for surfing, arrive sometimes days before the choppier surf generated by a storm.

WHAT DISTINGUISHES A ROGUE WAVE FROM A MERE BIG WAVE?[7]

As more and more instruments were deployed in the oceans in the 1990s, waves near to or greater than 100 feet high were consistently recorded.[8] During one storm in the North Atlantic, on Halloween 1991, the significant wave height[9] reached 39 feet, according to data from a buoy east of Cape Cod. One wave off the coast of Nova Scotia reached a height of 100.7 feet. The struggle of a fishing boat, the *Andrea Gail*, with these waves, and its likely demise by a rogue wave, became the basis for *The Perfect Storm*, a book by Sebastian Junger and a 2000 box office hit movie by Warner Brothers.

On New Year's Day 1995, a downward-pointing laser measuring device mounted on a drilling platform in the North Sea captured an enormous wave, now known as the "Draupner wave" after the name of the platform (Figure 7.4).[10] On most days, waves around the platform cause the ocean surface to rise and fall perhaps 10 feet as crests and troughs pass by. But on this day, the significant wave height was 36–40 feet. According to models at the time, the maximum height from trough to crest for these conditions would have been "only" about 66 feet, but a wave with a height of 86 feet passed under the laser. The probability of having a wave of this height under those sea conditions was about 1 percent—one way of saying that statistically only one wave of this height occurs every hundred years. The Draupner wave was truly a rogue.[11]

Rogue waves have several characteristics that distinguish them from more common and smaller waves. One characteristic, illus-

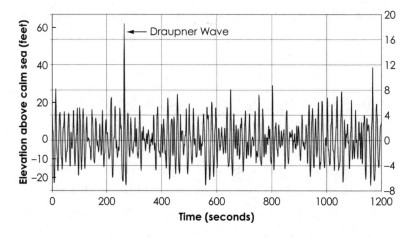

FIGURE 7.4 The record of the North Sea when the
rogue wave hit the Draupner platform at about 280
seconds on this record. *Data provided by Sverre Haver
of Statoil ASA.*

trated by the Draupner wave, is that the depth of the trough and
height of the crest do not match each other. The crest was 60
feet high, but the troughs on either side of it were only about 26
feet deep. The shape of the wave also is different: rogue waves
commonly have faces that are two to three times steeper than the
faces on the prevailing smaller waves at the time, giving them the
appearance of a nearly vertical wall of water as they approach.
Ship captain Ronald Warwick reported that when the *Queen
Elizabeth 2* collided with a 96-foot-high wave during a hurricane
in the North Atlantic in 1995, it looked as if the ship was heading
"straight for the white cliffs of Dover."[12]

In 2000, the European Space Agency (ESA) undertook project MaxWave to try to quantify the frequency and size of rogue
waves. Only a year after this effort began, two large tourist
boats, the *Bremen* and the *Caledonian Star*, were hit by rogue
waves at least 100 feet high. Over a period of three weeks
around this time, the satellites spotted ten waves higher than 80

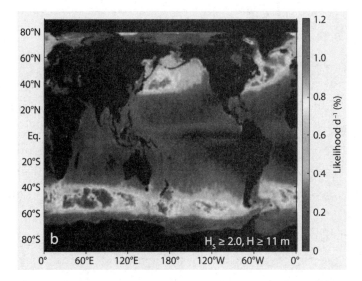

FIGURE 7.5 Map of the oceans showing the likelihood of encountering a rogue wave within any twenty-four-hour period. A rogue wave is defined here as one that has a height of twice the significant wave height, and a height exceeding 36 feet. The top of the scale represents, in percentages, the likelihood per day of a rogue wave—for example, 1.2 percent at the darkest colors. In this black-and-white version of the map, the most dangerous areas are the dark gray ones within the lighter shades, forming the two obvious bands in the two hemispheres. *From B. Baschek and J. Imai, "Rogue Wave Observations off the US West Coast," Oceanography 24, no. 2 (2011): 158–65.*

feet.[13] Later, an analysis of more than a million images covering the oceans showed that waves up to 100 feet tall were found most commonly in the North Atlantic, in the North Pacific, and in the Pacific Ocean southwest of Australia and near Cape Horn.[14] A separate study suggests that the average likelihood of encountering waves exceeding 36 feet in height along the main shipping routes in the North Atlantic is about 1 percent per day (Figure 7.5).[15]

SMOKE RINGS

Why do rogue waves occur in some parts of the world and not others? In the Southern Hemisphere they occur almost around the entire globe in latitudes between about 40° and 60°, making the South Atlantic and South Pacific dangerous for shipping between these latitudes (see Figure 7.5). They occur at roughly the same latitudes in the North Pacific, but they extend much farther to the north in the North Atlantic Ocean, from roughly North Carolina through Greenland and Iceland. The location of rogue waves in the Northern Hemisphere is influenced by the massive continents that disrupt ocean circulation there. In order to understand why rogue waves occur where they do, we need to delve into patterns of circulation of air and water on a global scale; that is, we get to dabble in meteorology and oceanography. These dabblings relate not only to rogue waves in this chapter, but also to discussions of weather, floods, and droughts in later chapters.

Ever since the fifteenth-century explorations for new trade routes, such as Columbus's voyages across the oceans, it has been known that surface winds in the latitudes from 0° to 30° in each hemisphere blow almost continuously from the east or northeast. These winds are called "trade winds" and are known as "easterlies" or "northeasterlies." The surface winds that blow from latitude 30° to 60°, the midlatitudes, blow from the west or southwest and are thus referred to as "westerlies" or "southwesterlies." In the northern polar region from 60° to 90°, the surface winds appear to come from the northeast and are called the "polar easterlies." There are similar patterns in the Southern Hemisphere, but the polar winds there are complicated by the presence of massive Antarctica.

Winds and currents form because the rate at which energy pours onto us from the sun is not uniform over the whole surface of the Earth—a fact recognized and incorporated into explanations for the trade winds as far back as the eighteenth century. The solar energy input is largest near the equator and smallest at the poles. Heat flows from warm to cold regions—that is, from the equator toward the poles. The atmosphere and the oceans work together to move heat around the planet away from the equator. This heat transfer creates the wind systems on our planet, and these, in turn, determine the near-surface circulation systems in the oceans. Ultimately, these processes determine how water is whipped into the frenzy that leads to the formation of rogue waves.

The simplest scheme for moving heat around in the atmosphere is one proposed by an English lawyer and amateur meteorologist, George Hadley, in 1735 (Figure 7.6).[16] In his model, a single "conveyor belt" moves air from the warm equatorial regions to the cold polar regions. Although the analogy isn't exact, imagine a gigantic distorted smoke ring blanketing each hemisphere from the equator to the pole, extending to about 10 miles in height. Warm air from the equatorial region rises, flows toward the poles along the top of the smoke ring, cools along the way, sinks back to the ground near the poles, and returns along the ground toward the equator (Figure 7.6). Heat is transported by the movement of air, the closed loops like this big one called "convection cells." The process is analogous to the way heat is transported in the "convection" setting for ovens. The motion of the atmosphere on Venus is somewhat like this simple single-cell scheme of convection, but unfortunately, this system is too simple for Earth.

Hadley knew that his model had a problem. In the model, winds in the tropical latitudes would blow in a north–south

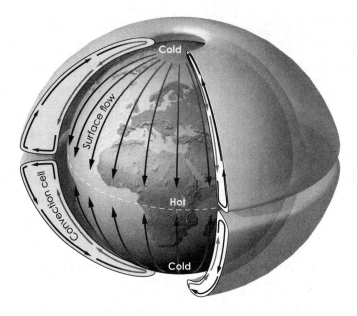

FIGURE 7.6 Circulation in a single cell as proposed by Hadley. Note that in the "giant smoke ring" analogy that I discuss in the text, the smoke ring wrapping itself around each hemisphere is smaller at the poles than at the equator.

direction as shown in Figure 7.6, but in reality they do not: they have both easterly and westerly components to their flow. Hadley attempted to explain this inconsistency by using the conservation of momentum, but he didn't get it quite right.[17]

For nearly a century, progress on the problem stalled as the likes of Kant, Laplace, Dove, and Foucault debated the issues. Finally, a hundred years after Hadley published his results, Gaspard-Gustave de Coriolis discovered a reason for the deviation from north–south paths,[18] and twenty-one years after that, William Ferrel[19] incorporated Coriolis's discoveries into a model for atmospheric circulation (published in a journal with the unlikely name of the *Nashville Journal of Medicine and Surgery*). Coriolis worked out how the paths of moving air are influenced by the rotation of Earth about its axis through the North and South Poles. Although this rotation is not strictly a "force" in the sense of classical physics and the conservation laws discussed in Chapter 2, its effect on the motion of bodies above Earth is often referred to as the "Coriolis force."

One way to think about wind and the Coriolis effect is to imagine wind as a river, and the river to be equivalent to a baseball being thrown from one place to another along a north–south line (a "meridian"). For example, take a wind that is flowing from the North Pole to the equator along a meridian of longitude. While the river of air is traveling across this hemisphere, Earth is rotating counterclockwise under it. Instead of staying on the meridian on which it was launched, the "baseball" of a river deviates to its own right as Earth rotates under it. In the Southern Hemisphere, it would deviate toward its own left.

Here's a second thought experiment to illustrate the Coriolis effect: Imagine a ruler laid down on a cardboard "record" on an old-fashioned LP turntable. Imagine that you are going to draw a line on the cardboard record by tracing a pencil down the ruler. If the turntable is not rotating, you will draw a straight line from the center of the record to its edge, and observers, both sitting on the record and sitting above it on a cloud, will see the same thing. If, however, the turntable is rotating slowly, the line on the paper will curve. Even though both lines were drawn along the straight ruler edge, they appear different on the cardboard because of the rotation of the turntable. The straight line mimics winds on a nonrotating Earth; the curved line, the way the winds curve on the rotating Earth.

Ferrel then built on the work of Coriolis and proposed, correctly, that heat transfer from the equator to the poles happens stepwise through not one, but three, major cells—one each very

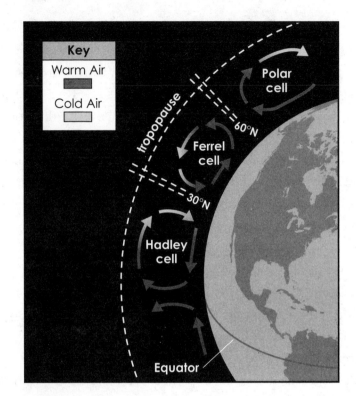

FIGURE 7.7 Ferrel's model of the circulation of air on Earth. Note that the vertical scale is greatly exaggerated. *Modified from http://science.howstuffworks.com/ nature/climate-weather/atmospheric/weather5.htm.*

roughly between latitudes 0° and 30°, 30° and 60°, and 60° and 90° (Figure 7.7). (These latitudes are highly variable, depending, for example, on the season of the year and how on much energy the sun actually emits, since its emission of radiation is not constant.) Each cell is very squashed vertically, extending only about 10 miles above Earth. From the equator to the poles, respectively, the cells are named the Hadley, Ferrel, and polar cells.

The boundaries between cells are, weatherwise, the most interesting places on the planet. Hot surface air rises into the two

Hadley cells at the equator, producing conditions there in which the pressure is generally low, the prevailing winds are calm, and skies can be alternately sunny, or cloudy and rainy, for long periods of time. Known technically as the "Intertropical Convergence Zone" (ITCZ), this region is the fabled "doldrums," described in Samuel Coleridge's *Rime of the Ancient Mariner*:

> *All in a hot and copper sky,*
> *The bloody Sun, at noon*
> *Right up above the mast did stand,*
> *No bigger than the Moon.*

> *Day after day, day after day,*
> *We stuck, nor breath nor motion*
> *As idle as a painted ship*
> *Upon a painted ocean.*

These conditions are occasionally interrupted by fierce storms that peel off the coast of Africa (Figure 7.8) and South America.

Once the warm air reaches the top of the Hadley cells, it spreads away from the equator, cools, and eventually sinks back down to the ground at about latitude 30°, where it returns toward the equator. At the bottom of the Ferrel cells adjacent to the Hadley cells, air moves north and south to about latitude 60°, where it rises and flows back toward the equator along the top of the cells. When it returns to about latitude 30°, it descends to the ground alongside the sinking air from the Hadley cells. The boundaries between the Hadley and Ferrel cells in each hemisphere are known as the "horse latitudes," regions of high pressure that receive little rain, and have generally clear skies and variable winds. In the Northern Hemisphere, the horse latitudes are roughly marked by the Sahara and the deserts of the southwestern US and northern

FIGURE 7.8 The doldrums, the equatorial boundary between the two Hadley cells, are marked in this NASA infrared image by the near-equatorial band of clouds starting from the northern part of South America and trailing west over the Pacific Ocean. At higher latitudes out in the Atlantic (extreme right), this image shows the remnants of Tropical Storm Claudette drenching the southeastern US, Tropical Depression Ana unwinding over Puerto Rico, and Hurricane Bill approaching from the central Atlantic. *NASA GOES 14 image, August 19, 2009.*

Mexico; in the Southern Hemisphere, by the Atacama, Kalahari, and Australian deserts (Figure 7.9).

The third boundaries, those between the Ferrel and polar cells, are referred to as the "polar fronts," and the one we know best is in the Northern Hemisphere. This polar front hangs out around the high latitudes during the summer, but in the winter it can dive down well into the midlatitudes in the southern US. It marks the position of the main jet stream. Here, cold polar air,

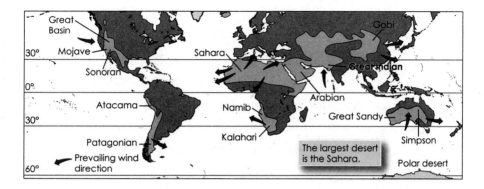

FIGURE 7.9 The deserts of the world. Notice how they cluster, very roughly, around 30° north and south of the equator, the boundary between the Hadley and Ferrel cells. *S. Marshak,* Earth: Portrait of a Planet *(New York: W. W. Norton, 2012).*

only slightly warmed during its journey south along the ground, rises adjacent to the rising air from the Ferrel cells, creating low-pressure conditions characterized by persistent cloudiness. Have you ever noticed the latitudes of some cities notorious for their gray weather? Helsinki (60°); Berlin (52°); London (51°); Moscow (55°); Stockholm (59°); Vancouver, Canada (49°); Seattle (47°); Reykjavik (64°); St. Petersburg (59°)? These are not just "high latitudes"; they are latitudes around the boundary between the Ferrel and northern polar cells.

FERREL CELLS AND SPINNING TOPS

The change in temperature from the warm (low-latitude) side to the cool (higher-latitude) side of Hadley, Ferrel, and polar cells is called the "temperature gradient." The temperature gradients in Hadley cells are fairly small: a traveler could traverse a Hadley cell in the Northern Hemisphere from south to north and the

weather would change only from tropical to subtropical conditions. Across a polar cell, it would change only from polar to sub-
polar climates. The temperature gradient in Ferrel cells is much
larger: a tourist traveling through the Northern Hemisphere Ferrel cell would go from the humid warmth of New Orleans almost
up to the freezing conditions at the Arctic Circle.

Convection works to reduce the temperature gradients by moving hot and cold air around. The task of transporting heat across
the Ferrel cells with their high temperature gradients yields dis-
tinctive weather patterns characteristic of these cells, including
the giant cyclones, anticyclones, hurricanes, and typhoons of the
midlatitudes. I like to imagine these weather systems as the result
of someone stirring the Ferrel cells from above with a bunch of
gigantic spoons. Each spoon points toward the ground, with its
handle sticking up 10 or 20 miles through the atmosphere. Every
once in a while, depending on the season, a few of the spoons
are stirred vigorously. The stirring sets a bunch of spinning tops
within the Ferrel cells into motion. The spinning tops are, for-
mally, vortices, and they range up to 600 miles in diameter. Just
as ball bearings lubricate motions in some mechanical systems,
vortices lubricate heat transfer in our atmosphere; the vortices are
the large-scale weather systems.

It is no coincidence that the largest number of Atlantic hurricanes occur between August and October, for it is then that
the contrast in temperatures between the still-warm ocean waters
in the Gulf of Mexico and the western Atlantic Ocean and the
encroaching cold Canadian air is strong. Both the "No-Name
Storm" of 1991 (the "perfect storm" of book and movie fame)
and Hurricane Sandy in 2012 occurred in late October.

Combining the concepts from these past few pages into one
illustration gives us one of the more complicated figures in this
book. Figure 7.10 summarizes the major wind patterns on the

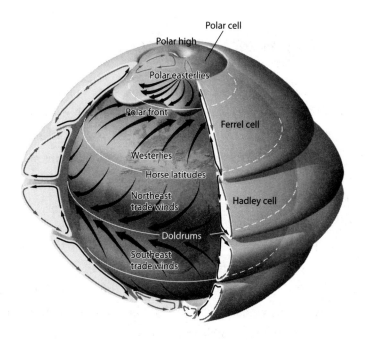

FIGURE 7.10 Generalized atmospheric circulation on
Earth according to the three-cell model, showing the
direction of winds near the ground and, in the cross
sections of the cells, their direction at the boundary
between the stratosphere and troposphere. *From
S. Marshak,* Earth: Portrait of a Planet *(New York:
W. W. Norton, 2012).*

whole planet: three cells in each hemisphere, surface winds
deflected by the Coriolis effect from the north–south trajectories
that would simply transfer heat from the equator to the poles on a
nonrotating planet, and boundaries between the cells with properties that depend on whether the air there is flowing up or down,
creating low- or high-pressure conditions, respectively. Now
compare Figure 7.10 with Figure 7.5, which shows the major locations of rogue waves, and ask yourself, "Where are rogue waves
most common?" The answer is, in the Ferrel cells.

OCEAN GYRES AND CURRENTS

The presence of continents on Earth causes the wind patterns to be more complex than predicted by the three-cell model (Figure 7.11*a*). In particular, patterns of air circulation around high-pressure cores develop in the Northern and Southern Hemispheres over the oceans. The winds, in turn, drive the near-surface waters on the ocean to produce currents and waves of all sizes, including rogue waves.

Most pronounced are the huge rivers within the ocean—surface currents that extend down to a depth of about 1,000–3,000 feet.[20] These currents flow in roughly circular patterns called "gyres" (rhymes with "tires"). There are five major gyres: in the northern Pacific, northern Atlantic, southern Pacific, southern Atlantic, and Indian Oceans (Figure 7.11*b*). In the Northern Hemisphere the gyres have clockwise circulation; in the Southern Hemisphere, counterclockwise. Very roughly, the gyres seem to rotate in patterns that mimic wind patterns (compare Figures 7.11*a* and *b*), but because water is so much heavier than air, its response to the Coriolis effect is much slower. As a result, the Coriolis deflection of the water is offset from the wind direction by as much as 45 degrees.

The surface currents in the ocean are strongly influenced by the presence of the continents, but the major ones are quite obvious if you simplify their pattern in your mind to be a rectangle. Take, for example, the North Atlantic Gyre. The westerly winds between latitudes 30° and 60° push the water eastward, forming the North Atlantic Current, or North Atlantic Drift (see Figure 7.11). The northeast trade winds near but north of the equator push water there to the west, forming the North Equatorial Current. These two east–west currents are con-

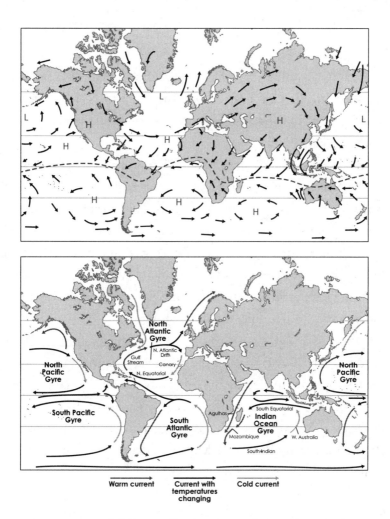

FIGURE 7.11 (a) Generalized surface wind patterns typical of January and, for comparison, (b) the five major
gyres of ocean circulation. In (a) high- (H) and low- (L)
pressure regions of the atmosphere are shown by the
solid contours, and rotating wind patterns by the arrows.
*Part (a): modified from the COMET® website at http://
meted.ucar.edu of the University Corporation for Atmospheric Research (UCAR), sponsored in part through
cooperative agreement(s) with the National Oceanic and
Atmospheric Administration (NOAA), US Department of
Commerce (DOC). ©1997–2013 University Corporation for
Atmospheric Research. All rights reserved. Part (b): modified from M. Pidwirn, "Surface and Subsurface Ocean
Currents: Ocean Current Map" in* Fundamentals of Physical Geography, *2nd edition (2006), as found at http://
www.physicalgeography.net/fundamentals/8q_1.html.*

nected by two other currents in the north–south direction: the Gulf Stream flowing north along the coast of North America, and the Canary Current flowing to the south on the eastern side of the North Atlantic.

Or take the Indian Ocean Gyre. The Indian Ocean is mainly in the Southern Hemisphere. The southeast trade winds push water to the west, forming the South Equatorial Current. On the opposite side, westerlies move the water eastward, into the South Indian Current. The gyre, rotating counterclockwise in the Southern Hemisphere, is completed by the West Australian Current moving north, and the Mozambique and Agulhas Currents, moving southward along the coast of Africa.

Most rogue waves are formed within the major currents (compare Figures 7.5 and 7.11). In the northern Atlantic they occur in the Gulf Stream (including the infamous Bermuda Triangle), and in the East Greenland and Norwegian Currents. In the northern Pacific they are in the Kuroshio Current; and in the southern Pacific, in the Agulhas and Mozambique Currents on the east coast and tip of southern Africa.

THE REST OF THE STORY

Although these broad patterns of wind and water movement determine the regions where big waves occur, other factors go into producing a rogue wave. In particular, the generation of rogue waves depends on several processes working simultaneously,[21] including the churning up of the ocean surface by winds from hurricanes and storms that can be quite far from the rogue wave zones; the piling up of waves from the deep ocean into shallow depths along the continental shelves; the interaction of strong waves moving in opposite directions, such as storm waves interacting with strong oceanic currents or strong opposing winds; and the constructive interference (addition of wave heights) of random waves. Let's look at each of these briefly.

Imagine the quiet, flat sea of the becalmed ancient mariner.[22] When a wind arises, the air quickly becomes turbulent and gusty; that is, it has swirls and pockets moving at different velocities and in different directions—an effect easily observed in the autumn when wind picks up leaves on the ground. The swirls and pockets of air produce pressure fluctuations on the sea surface that generate small waves—say, a few inches in wavelength. These small waves poke up out of the formerly smooth surface of the water, providing little knobs, so to speak, that can be pushed by the wind to generate bigger waves; that is, small waves grow. The small waves interact with each other, and big waves gobble up small waves.

Waves—in particular, rogue waves—don't form immediately when a wind starts to blow. Rather, wind needs to blow for a fairly long time across a long and open stretch of water before waves are fully developed. For example, a 45-mile-per-hour wind blowing over a large area of the North Atlantic takes several days to fully develop the strong waves.

The second effect—piling up of waves in shallow water, or "shoaling"—is so similar to the dynamics of tsunamis coming onshore (discussed in Chapter 6) that I won't spend time on it here, other than to point out that a large concentration of shipping accidents happens along the continental shelves, where water gets shallower and shallower toward the coast. Shoaling plays a major role in these events.

The third effect—the interaction of waves moving in opposing directions—is well illustrated by shipping accidents along the east coast of South Africa,[23] where waves in the Agulhas Current can be up to 100 feet high and a disproportionate number of ships have sunk. Rogue waves in this region claimed the *Waratah*, a 456-foot-long passenger/freight ship, in 1909; the *World Glory*, a 736-foot-long tanker, in 1968; and the *Neptune Sapphire*, a cargo ship on its maiden voyage, in 1973. Since 1990, at least twenty ships have been hit by rogue waves in the Agulhas Current. Intriguingly, this was an area avoided by early Arab sailors because of their knowledge of the sea, and it is suspected that this avoidance contributed to the fact that, as far as we know, Portuguese navigators, not Arabs, became the first to round the Cape of Good Hope in the expedition of Bartolomeu Dias in 1488.[24]

This area of the ocean lies between latitudes −40° and −60° south, where, unlike conditions at the same latitudes in the Northern Hemisphere, no continental-scale landmasses block the winds, so they can blow for enormous distances, creating perfect conditions for winds and currents to produce huge rogue waves. In the winters (June–August), strong low-pressure systems develop south near Antarctica.[25] As these storms move to the northeast, they are preceded by winds blowing in from the northeast—that is, winds blowing in the same direction as the Agulhas Current flows. Such winds push on the current, whipping up its speed. After the storms pass, however, the winds switch and blow from the opposite direction, from the southwest. These southwesterly winds are so strong that they would create waves 20–40 feet high even in the absence of the Agulhas Current. But when these waves meet waves in the strong opposing current, the two can nearly be stopped in their tracks. The waves steepen and, potentially, become rogue waves.

This situation in the Agulhas Current is an example of a fourth effect contributing to the formation of big waves—constructive interference, waves meeting each other with their heights adding together. Constructive interference produces large waves, but they are often very short-lived. This effect is particularly pronounced near coastlines, where landmasses reflect waves in a very complicated way. The rapid times over which they form and disappear make them dangerous to the shipping industry and difficult to document scientifically.

In 2007 there were nearly 35,000 commercial vessels with gross tonnage of more than 1,000 tons, and more than 4 million small to large fishing vessels, used by 30 million fishermen. Rogue waves on the ocean or other large bodies of water are dangerous to these vessels, to oil and gas platforms, and even, as we saw at the start of this chapter, to rescue helicopters that must fly close to the surface of the water. As we search more and more in the seas and oceans for food and resources, ever more people and equipment will be exposed to risks from ocean storms and rogue waves.

REFLECTIONS: ROGUE WAVES, OPTICAL FIBERS, AND SUPERFLUID HELIUM

Although the loss of ships and ocean platforms has decreased over the past century because of technological advances, the rate of loss is still high. Advances in weather monitoring and weather forecasting in the last few decades have greatly improved the information available to the shipping industry. Nevertheless, rogue waves are both spatially and temporally rare, which causes scientists great difficulty in studying their characteristics, developing models for their dynamics, and testing the models in actual rogue wave settings.

Two aspects of description of these dynamics are made especially difficult by the rareness of the waves—their general properties, such as the shape of the waves, and their statistical properties, such as their frequency of occurrence in any particular condition of oceanic disturbance. The equations that describe these phenomena are daunting even to most scientists, and the chances of testing the conclusions derived from such equations with oceanographic data on rogue waves are very small.

In the Reflections section of Chapter 3, we looked at how scientists try to combine their fundamental understanding of earthquakes with the tools of statistics and probability to help in forecasting hazards, and in Chapter 6 we saw how historical records combined with geologic data can provide models. Fortunately, there is yet another method by which scientists study phenomena—by finding analogues in other settings that give a complementary understanding. If we can find analogues that display similar dynamics and similar statistics to those of rogue waves, study of those systems can be applied to supplement the understanding that we gain from actual observations of the waves themselves.

In fact, the phenomenon of rogue waves may prove to be quite universal, with suggestions that they exist not only in oceans but in the atmosphere, in optics, plasmas, superfluids, and capillary waves.[26] In one laboratory setting, scientists produced waves that have similar properties and obey similar equations by running laser light through special optical fibers. When a light beam having a very specific and narrow wavelength is put through an optical device that spreads that wavelength out over a very broad range, rare bright flashes of light, so-called optical rogue waves, are generated. In another laboratory setting, waves with similar, but not identical, properties can be produced in superfluid helium (technically,^{4}He), an anomalous and complex state of helium that exists at temperatures very close to absolute zero. In the experiments that produce the waves, heat applied to the system is transferred by waves, analogous to pressure waves like those from a bursting bicycle tire. These heat waves, referred to as "second sound" by their analogy to normal sound waves, develop spikes (the rogue waves) under certain laboratory conditions. Rogue waves in these laboratory settings, while rare in the context of the laboratory experiments, actually occur quite frequently compared to rogue waves on the ocean because the systems producing them are small.

Efforts to explain and model the rogue waves in each of the three systems—ocean, optical, and superfluid helium—lead to sets of equations which, to first order, are quite amazingly similar to each other. Because the settings of the laboratory experiments can be controlled and sampled with sophisticated instruments, predictions of the theories can be tested in the laboratory and then extrapolated to the larger oceanic rogue waves. This is an active field of research that has only emerged in the twenty-first century.

The existence of rogue waves in the atmosphere is still, as of this writing, a matter of speculation. However, there is no doubt that mysterious waves appear in the atmosphere, some that pose serious danger to private and commercial aircraft, some of which may eventually be proven to obey the same dynamics as rogue waves. In the next chapter, we explore some of the characteristics of these waves and existing explanations for their properties.

Chapter 8

RIVERS IN
THE SKY

THE GLASS HOUSE, JOPLIN, AND CHOPPING ICE "FOR CULINARY PURPOSES"

Winds are a problem not only at sea, where they drive the rogue waves discussed in the previous chapter, but on land as well, where they cause much of the damage in storms such as Hurricanes Katrina (2005) and Sandy (2012). Few scientists can describe winds as well as Raymond Chandler did in his 1938 short story "Red Wind":[1]

> There was a desert wind blowing that night. It was one of those hot dry Santa Anas that come down through the mountain passes and curl your hair and make your nerves jump and your skin itch. On nights like that every booze party ends in a fight. Meek little wives feel the edge of the carving knife and study their husbands' necks. Anything can happen. You can even get a full glass of beer at a cocktail lounge.

I experienced such mountain pass winds once when I lived in a two-story house perched on a cliff in Howe Sound, a beautiful 40-mile-long fjord that opens into the Strait of Georgia on the West Coast near Vancouver, Canada. With floor-to-ceiling glass windows prone to rattling so loudly that I thought they might implode any minute, the strong winds provided acoustic percussion to my spectacular view—sometimes, in December and January, for four to five days at a time.

Inland (east) of the coastal mountains that border Howe Sound lies the large, high, interior plateau of central British Columbia, and to the west lie gateways to the Pacific—the Straits of Georgia and Juan de Fuca. During winters, cold polar air from the interior is funneled through the mountains toward the straits down fjords like Howe Sound.[2] These events are accompanied by extreme cold, with wind gusts reaching 65–90 mph. Trees toppling in the large fir forests commonly cause very serious power outages during the coldest periods of the bitter Canadian winter.

Just a few years before I arrived in British Columbia, an outbreak of Arctic air in late January–early February 1989, known as "The Big Chill" or "The Alaska Blaster,"[3] brought record low temperatures to much of western North America. Strong winds blowing through Howe Sound knocked trees into power lines, leaving 20,000 homes without electricity, some for several days in the freezing weather. A water reservoir froze, leaving 70,000 people without water. Records were set for electricity and natural gas consumption. Ferry service between Vancouver Island and the mainland was curtailed, and heavy ice from freezing sea spray damaged boats and properties along the sound.

Having moved from southern Arizona, I wondered, "What

have I gotten myself into living on this cliff with only a wall of glass separating me from the violent world of winds outside?" While I lived in this house, I was unaware of the uniqueness of the meteorology of the straits, and only much later found this description, written in 1931:[4]

> *The easterly gales at the west end of the Strait of Juan de Fuca constitute one of the notable climatic eccentricities of the North American Continent. Indeed it may not be extravagant to claim for them a position unique among the winds of the world. The writer knows nothing in meteorological literature which describes their counterpart, although winds of similar type though less violent are common to many other localities.*

Years later, I moved to central Illinois, in the heart of America's "Tornado Alley," and learned what truly strong winds are. Midwesterners live in a different relationship to glass windows than do residents of British Columbia, with some having completely windowless underground bunkers to take shelter from tornadoes. On a wind intensity scale of 1–4, my winds in Howe Sound were a paltry "1" with winds possible up to 130 mph (Figure 8.1). Tornado Alley, on the other hand, rates a "4" with possible winds up to 250 mph.

The difference in wind behavior between Howe Sound and the Midwest was tragically illustrated when a ¾-mile-wide tornado blasted through Joplin, Missouri, around dinnertime on May 22, 2011. On a tornado intensity scale of 1–5, this was a 5, with winds exceeding 200 mph. Everyone who lives in Tornado Alley knows that the first response to a tornado alert is to move away from windows, into a tornado shelter, basement, interior room, or even

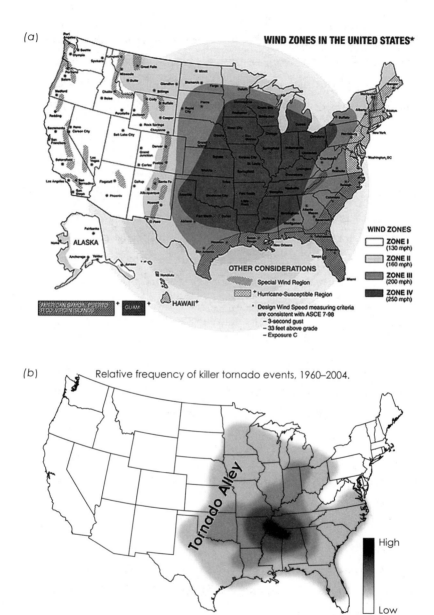

FIGURE 8.1 Wind (a) and tornado (b) zones in the US.
FEMA.

a hallway, in order to get away from flying glass and other debris. In St. John's Regional Medical Center in Joplin, the warning "Execute condition gray!" went out. Nursing staff immediately started a well-rehearsed drill to wheel the beds of patients into hallways. Fortunately, most patients had been moved by the time the tornado hit, but windows exploded and IV lines were ripped from patients' arms.

Nearly every patient was bloodied, either from flying glass or from being splattered by blood because of being in proximity to someone else who was bloodied.[5] Critical medical equipment was left without power. People in the emergency room were sucked out of windows into the parking lot. One patient in the emergency room for treatment of a single broken rib ended up with two more broken after the storm. It was the deadliest tornado in more than sixty years, killing 153 people and leaving nearly 1,000 with injuries.[6] Five patients in St. John's hospital died as a result of the storm. Others, including some newly injured by the storm, were sent to hospitals around the state, leaving family members frantically driving from one hospital to another to locate missing relatives. Unexpectedly, nearly a dozen people developed a fungal infection, zygomycosis, from soil or plant material that had penetrated their skin. This fairly rare infection can be cured only with harsh measures: strong antifungal drugs and physical removal of damaged skin, some of which, in this case, had obvious mold in the wounds. At least three or four of the 153 deaths were caused by this infection.

Strong winds are not unique to the Midwest. The strongest winds on the planet blow nine months of the year in the Antarctic (Figure 8.2). The Antarctic explorer Ernest Shackleton described winds near McMurdo Sound:[7]

FIGURE 8.2 A member of the Mawson expedition to the Antarctic leaning into a steady 100-mph-wind while picking ice "for culinary purposes." Titled *Leaning on the Wind*, the photo was taken sometime between 1907 and 1914. *From "With Shackleton to the Antarctic"; photo by Captain Frank Hurley.*

Occasionally on this night, as we approached the eastern shore, the coast of Ross Island, we noticed the sea covered with a thick yellowish-brown scum. This was due to the immense masses of snow blown off the mountain sides out to sea, and this scum, to a certain extent, prevented the tops of the waves from breaking.

The dynamic condition of the atmosphere that envelops Earth is one of the most visible examples of changes of state. To explore the causes of winds and the factors that influence their strength, our field trip ranges northward from the Antarctic through California to British Columbia, and then back southward to Joplin.

PRIMER: RIVERS OF WATER

In one part of my career, I studied a "killer wave" in one of the major rapids of the Colorado River in the Grand Canyon. As a result, I have had a love affair with the Colorado and its rapids for more than forty years, and have studied in detail why the rapids exist.[8] Much to my amazement, I was to discover, and will explain below, that a counterpart of the Colorado—a "river" of air in the sky—had its own "rapids," and its waves were washing back and forth across my windows in Howe Sound.

The way that water flows in a river is fascinating, diverse, and complex. In the Midwest, rivers like the Mississippi typically wend their way inexorably toward the sea as a solid, nearly wave-less mass of water, hosting great fleets of massive ships carrying wares into and out of the interior of the continent. In contrast, in the mountainous areas of the West, rivers like the Colorado almost dance their way to the sea—rushing headlong across the landscape, across small cliffs and rocks, making foaming, dancing stretches known as "rapids"—areas of great challenge to kayakers and rafters, but of great hindrance to bigger boats or commercial transport.

These two river styles are fundamentally different, not only to commercial shipping and recreational sports, but to scientists who study them. The hydraulic phenomenon that causes the difference can be illustrated by a simple experiment that you can do in your kitchen sink (Figure 8.3*a*). Place a flat plate on the bottom of the sink, and turn on the water, adjusting its flow until you see a pronounced circular wave on the plate. Where the downgoing jet hits the sink, water spreads out very fast and shallow until it reaches a cylindrical "wall." Beyond this wall, the water still flows away from the jet, but it is deeper and slower. The wall separates two regions of flow with very different properties—fast and shallow versus slow and deep. The wall is called a "hydraulic jump." It has a cylindrical geometry on the plate because it is created by a cylindrical jet, but you can see similar "jumps" in water flowing down streets during rainstorms, in the big waves that form at the spillways of dams, in water flowing over submerged structures such as weirs, in many experiments you could do with a garden hose, and in many rivers (including those in the Midwest if you look closely at, for example, shallow flow near the shoreline).

The hydraulic jump separates two regimes of water flow with very different properties (see "Flowing Rivers: Changes in Regime" and Figure 2.4 in Chapter 2). These flow regimes have many properties analogous to "subsonic" and "supersonic" airflow, and engineers use similar words—"subcritical" and "supercritical"—to describe these two flow regimes in water. The water changes from the supercritical (fast and shallow) regime to the subcritical (deep and slow) regime by a sudden deceleration as it passes through the hydraulic jump. For big (15–20 feet high) waves on rivers, the change in velocity at the hydraulic jump can be 20–30 mph. Sometimes in rivers, after water goes through one hydraulic jump it is still flowing downhill and can accelerate again, causing another hydraulic jump to form downstream of the first, and so on. These hydraulic jumps form the waves in rapids that are such a thrill to river kayakers and rafters. However, hydraulic jumps have a side so dangerous that they have been called "drowning machines"[9] (Figure 8.3b). They can be highly turbulent, with water moving violently in all directions, and they often contain water spinning around a horizontal axis—a so-called horizontal vortex that can trap unwary swimmers and rafters.

(a)

(b)

FIGURE 8.3 (a) A hydraulic jump separates two regions of different flow characteristics. (b) A sign posted near a low-head dam to warn people to stay away, illustrating the danger of being sucked under into the rotating, horizontal vortex. *Part (a): courtesy of John Bush; part (b): courtesy of anonymous photographer through www.geocaching.com.*

RIVERS OF AIR

The low-level winds in our atmosphere are rivers of air that travel in the "troposphere," which is also where we live. The troposphere extends up to about 30,000 feet at the poles and to about 56,000 feet at the equator. In the atmosphere, the average pressure decreases with altitude. (Travelers often encounter this phenomenon when they pack a bottle of lotion, for example, at sea level and then open it in a city, such as Denver, at higher altitude. An unexpected spray often erupts from the bottle, driven by the sea-level pressure stored inside the bottle.) The average density and temperature of the atmosphere also decrease with altitude in the troposphere. The temperature behavior is much more complicated at higher altitudes.

Because the surface of Earth is not flat, with some parts sticking up higher into the atmosphere than others, pressure, temperature, and density vary along the surface just because of this structure of the atmosphere. Locally, however, pressure, temperature, and density are always changing, and deviating from the averages—by the minute, hour, day, month, and season.

At the scale of fairly large regions, weather is controlled by the conditions of air over the oceans and landmasses. Climate and weather[10] along a coast are always moderated by the ocean, with high and low temperatures never being as extreme as those inland, where weather can be very hot during the daytime or in the summer, and very cold during the night or in the winter. As a result, coastal air masses can have either higher density and pressure than inland air masses have (daytime or summer), or lower density and pressure (nighttime or winter). During hot weather, winds tend to blow from the higher-pressure coastal regions into the warmer lower-pressure inland areas; these are the so-called

FIGURE 8.4 A likely hydraulic jump in the atmosphere,
the sharp linear feature running from left to right
through the top of the photograph, where a fast atmospheric flow coming south from Canada met quiet
slower air over the Great Lakes. A time-lapse movie of
this is available at http://cimss.ssec.wisc.edu/goes/blog/
archives/652. NASA GOES 12 image.

sea breezes. During cold weather, conditions are reversed. Large
systems of high and low pressure form in response to differences
of temperature and humidity, and these pressure differences also
create winds. The size of these systems varies greatly, from 10 to
125 miles (the scale of so-called mesocyclones) or from 125 to
1,250 miles (the scale of tropical cyclones).

Several factors drive winds, or influence their intensity, on
smaller regional scales. One is the flow regime already discussed.
Four additional factors (which I will discuss) also drive winds:
the density of the air, the force of gravity, the shape of the land
(topography), and pressure gradients (how rapidly pressure
changes with distance). Generally, more than one force is operat-

ing, and often all four play a role simultaneously in the complex meteorology of the planet.

Once air is set into motion, a fifth factor comes into play that can alter the direction of the winds: the Coriolis effect discussed in the previous chapter. In response to all of these factors, the flowing air enters one or the other of the two hydraulic regimes (subcritical and supercritical) described earlier (Figure 8.4). All of these effects combine to yield a very complex set of wind patterns at scales ranging from a mile (tornadoes) to continental (the biggest pressure systems). It would take a textbook in meteorology (as well as engineering hydraulics) to cover all of the possibilities, but we can look at the roles of density, gravity, and topography in some of the most interesting, and most relevant, settings to our daily lives.

WINDS THAT FLOW *OVER* TOPOGRAPHY: FOEHNS, CHINOOKS, AND STEVE FOSSETT

Dense, cold air originating high on plateaus in the continental interior or over an icy polar cap reacts to the force of gravity by racing down to the lower coastlines, sometimes reaching hurricane force.[11] These dense, cold winds are rivers of air that flow as discrete layers sliding under the warmer coastal air. Shackleton estimated that the winds like this in the Antarctic had speeds of 100 mph; winds of 200 mph have been recorded there by modern instrumentation.

This cold air warms as it flows downhill toward the coasts because it is being compressed to higher pressure—a process not unlike that when you fill your car or bicycle tires with air. The energy for the warming in a tire comes from your work at the pump, but in nature it comes from the potential energy of the air on the high plateaus as it is converted into kinetic energy and heat while flowing downhill. Often this compression is cited as the only reason for

FIGURE 8.5 The McMurdo Dry Valleys of the Antarctic, snow-free because of the hot, dry winds that roar down from the interior mountains. *Image by NASA/GSFC/METI/ ERSDAC/JAROS, and US/Japan ASTER Science Team.*

warming at the base of lee slopes of mountains, but another effect that can be as large, or larger, is that the river of warm air displaces pools of cooler air that might lie at the base of the mountains.

In spite of the compression of air along its way to the coast, winds in the Antarctic remain bitterly cold as they travel, and often they are still bitterly cold when they arrive. Along the way, snow is vaporized (sublimated) and scoured away by the force of the winds, some being taken to the ocean to become the "scum" described by Shackleton. Several valleys near the US research base at McMurdo Sound are scoured completely free of snow, forming unique conditions that give them the name "McMurdo Dry Valleys" (Figure 8.5). The valley floors are covered not with ice and snow, but with loose, dry gravel forming a stark, desiccated landscape. (Yet in spite of the dry and cold conditions, bacteria have been found in the dry valleys within "moist" rocks, wetted by summer meltwater from the glaciers. The Dry Valleys may be the closest place on Earth to the surface of Mars.)

Winds driven by gravity also occur in North America, perhaps the most notorious being the "Santa Anas " in southern California, the winds described by Raymond Chandler in "Red Wind." The Santa Anas form when air flows westward to the Pacific coast from the Great Basin of Nevada. In winter, the Santa Anas can blow cold and dry, often bringing some of the coldest weather of the year to southern California. More dangerously, in the autumn after a wet spring that has nurtured the growth of vegetation, the Santa Anas, blowing hot and dry, can fan serious brush and forest fires (Figure 8.6).

What is it like to live with such winds? As might be guessed from Chandler's passage, these winds have been accused of causing a variety of maladies, particularly migraine headaches.[12] There are physical dangers as well. Whereas in Joplin it took actual penetration of the skin of victims to expose them to fungal infections, merely breathing during the Santa Anas can be dangerous. The winds carry a pathogenic fungal spore that causes an influenza-like condition called valley fever (coccidioidomycosis). Although generally not serious, it can progress into skin ulcers, bone lesions, and general inflammation problems.

When winds flow from the coast over mountain ranges on their way to the continental interior—for example, from the Pacific Ocean over any of the western coastal mountain ranges, like the Sierra Nevada—a variety of interesting phenomena occur. The air starting at the ocean is moist. As this air cools, the moisture can condense to form rain or snow on the windward side. This process produces the notorious wet, gray weather in many

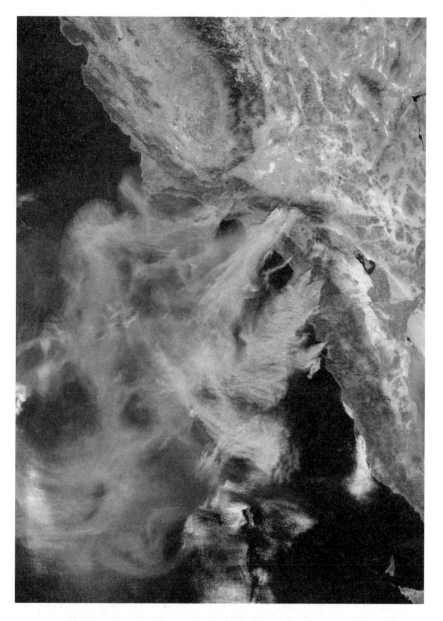

FIGURE 8.6 Santa Ana winds in southern California
blowing dust and smoke from the wildfires of October
25–26, 2003. *Image from NASA.*

regions west of the mountains in the western parts of the US and Canada.

The formation of rain or snow on the windward side of a mountain range has two consequences: the rising air dries out and also warms up because condensation of water vapor into rain droplets releases heat (called "latent heat"). (Many people have discovered latent heat the painful way—by accidentally passing a hand or arm through the plume of steam emanating from the nozzle of a teapot. As steam condenses on the skin, it gives up the latent heat, causing a skin burn.)

Air flowing over a mountain range reaches its lowest temperature at the crest. Once over the crest, the air, now much drier than when it started its journey on the windward side but still possibly containing some moisture, picks up downhill speed. It also warms because it is compressed as it descends. The warming can be substantial—so substantial that it produces hot, dry winds. Just as the winds blowing out to sea are given names like the Santa Anas, notorious winds on the leeward side of mountains are often given names. These depend on the specific geographic region, but two famous ones are the "foehn" (a European term) and the "chinook" (an American term in the Rocky Mountains). Appropriately, since these winds can sublimate and scour snow from the surface, "chinook" is a Blackhawk Indian term for "snow eater."

As the air flows down the slopes of mountains, many different flow configurations can occur, often lumped together under the broad term "lee waves." Hydraulic jumps may or may not occur. They can develop on mountain slopes or out in the flatter lands at the base of mountains. There can also be a horizontal column of spinning air—a vortex—with its axis parallel to the mountain front, like the one illustrated in the drowning machine sign of Figure 8.3b. Because of the rotation of the vortex, these features are appropriately named "rotors." They are regions of notorious and dangerous turbulence.[13]

Unlike hydraulic jumps in water, rotors aren't usually visible, making them extremely hazardous for air navigation. The leading edge of the rotor is nearly vertical, forming a straight barrier sometimes extending crosswind the whole length of the mountain range. The rotor, and clouds that cap it, can reach up to 25,000–30,000 feet, higher than the normal cloud deck over the crest of the mountain range.[14] Just as rafts floating in river rapids "hit a wall" when they encounter a hydraulic jump, aircraft "hit a wall" when they encounter a rotor. One sailplane intended for investigating the properties of rotors was designed for accelerations or decelerations of 8–10 g, but it appeared to have undergone 16 g before being destroyed. (For reference, an elevator accelerates upward at a g-force of about 1.14 g. Fighter pilots are trained to withstand a g-force of 9.)

Rotors are one of the most attractive settings for aviation adventurers. Gliders have used these waves to ascend to great altitudes ever since rotors were discovered by German glider pilots in the 1930s. Einar Enevoldson and the late Steve Fossett ascended to the current world record of 50,727 feet using the dynamics of rotors.[15]

However, rotors are inherently dangerous. One day in 1964, sailplane pilots in the vicinity of the Sierra Nevada in California reported having visual evidence of a "tremendous" wave in the lee of Mount Rose, Nevada.[16] That same day, about 20 miles away, a Paradise Airlines plane crashed into the mountains while on a scheduled flight from Oakland to Lake Tahoe Airport, killing all eighty-five people aboard. The cause was apparently turbulence in a rotor.

Two years later, a British Overseas Airways Corporation plane (BOAC Flight 911) taking off from the airport in Tokyo crashed downwind from Mount Fuji, only 40 miles south of Tokyo, killing 124 people. Mount Fuji is a beautiful mountain, but notorious for its high winds and tricky air currents. On that day, winds at the summit were 70–80 mph, the weather was clear, and there were no clouds to provide markers of the presence of waves. The pilot of Flight 911 filed for permission to climb via a route that would take them closer than normal to Mount Fuji, possibly to treat the passengers to a beautiful view. Approaching the mountain from the downwind side, he flew into the invisible lee waves and, probably, a rotor.[17] The plane broke up and disappeared, leaving a 10-mile-long trail of wreckage. A US Navy A-4 Skyhawk sent out to search for wreckage encountered severe turbulence with g-force accelerations from +9 to −4 on an accelerometer scale.

In perhaps the most famous accident due to rotors, a chartered plane carrying the 1972 Uruguayan rugby team encountered a rotor at an altitude of 12,000 feet in the Andes. The pilot lost control, the aircraft crashed, and twenty-nine of the forty passengers died. Two survivors were able to hike to civilization, but the remaining survivors were not rescued for over two months. Their survival, and the role that cannibalism played, formed the basis for the book and 1993 movie *Alive*.

WINDS THAT FLOW *THROUGH* TOPOGRAPHY: GAP WINDS

> It was one of those hot dry Santa Anas that *come down through the mountain passes*.
>
> —Raymond Chandler, "Red Wind" (emphasis added)

Although winds respond to gravity, they are often channeled into paths that twist and bend through mountain passes and valleys in the topography. Even urban dwellers can appreciate the fact that topography plays a role in wind strength, because "street canyon winds" that funnel through gaps between buildings in cities can be very strong. New Yorkers are familiar with a frigid winter wind that can nearly knock them off their feet as it roars down Thirty-Third Street near Penn Station. In nature, as well as in urban settings, these winds are especially strong when atmospheric storms drive winds in the same direction as gaps and valleys in the topography. These are the winds that pounded my glass house in Howe Sound, known there also as "Squamish winds."

Rarely do so few letters of the alphabet provide such a great and concise description of a phenomenon as the word "gap" in the term "gap wind" that has been applied to such winds (Figure 8.7a). Introduced in 1931[18] to describe winds in the region of my glass house, the term has become a generic name for strong, low-level winds between two mountain ranges or a gap in a mountain barrier.[19] Winds such as these occur in the Antarctic and in other parts of the world as well, perhaps one of the most famous being the "mistral," which flows through the Rhône valley in France. Provence, the popular tourist destination and wine-growing country in southeastern France, owes its climate, particularly the crystal clear air and incredible number of sunny days compared to the rest of the country, to the mistral.

The river of cold air flowing seaward through gaps is often less than a half mile thick, and it flows as a well-defined layer through low-lying valleys and mountain passes (Figure 8.7 b). Often the valleys are bounded by steep mountains and have tortuous shapes, resembling nozzles with many contractions and

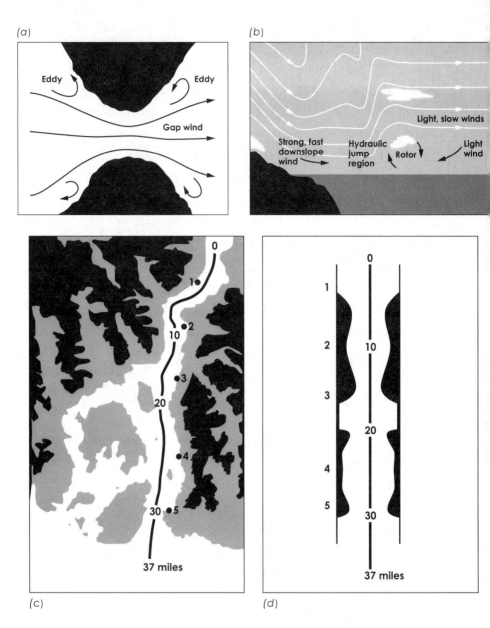

FIGURE 8.7 (a) A simple gap wind. (b) A lee wave with hydraulic jump. (c) A tortuous natural nozzle. (d) A simplified laboratory
model of the geometry in (c). In (c) and (d) the solid line extends
from 0 to about 40 miles. My house was at point 5 on the lower right
of part (c). An excellent resource for understanding the gap winds
in this area and for part (a) of this figure, can be found at http://
www.islandnet.com/~see/weather/elements/gapwind.htm.

expansions (Figure 8.7*c*,*d*). Nozzles have the same effects on flow in the atmosphere as they do on the flow of water in river channels: air accelerates into the nozzle, passes through a constriction, and exits the other end of the nozzle, sometimes through standing waves and hydraulic jumps. Scientists flying in research aircraft, interpreting data from wind monitoring stations, and using numerical and physical models are now able to measure and describe conditions in these rivers of air. Near my house in Howe Sound, they found regions of subcritical and supercritical flow, as well as not one, but two, hydraulic jumps separated by about 10 miles.[20] The biggest hydraulic jump was located near my glass house! Hydraulic jumps in the atmosphere move around compared to those in rivers because the speed of wind is so variable. So, my windows were being attacked not only by high winds, but by a big and turbulent hydraulic jump, and possibly even a rotor, washing back and forth across them.

One of the most disconcerting features of the Squamish winds beating on my house was the alternation between periods of very high velocities and curious silent lulls. Similar lulls in the fierce winds were noted in the early part of the twentieth century by observers around the coasts of Antarctica.[21] These changes in wind velocity and wind direction are the most dangerous aspects of the gap winds for small watercraft because the winds often blow out of the canyons for some distance over the water. If the Antarctic explorer in Figure 8.2 had been in a gap-wind setting instead of a constant 100-mile-per-hour wind, he would have had little chance of chopping out ice for his "culinary purposes," because he'd likely have fallen on his face when a lull occurred!

THE BIGGEST RIVERS IN THE SKY:
OUR JET STREAMS AND
HURRICANE SANDY

The largest-scale winds on the planet are found in our jet streams. There are two major jet streams in each hemisphere: the polar and subtropical jets (Figure 8.8). Both played a role in producing and steering the monstrous Hurricane Sandy to landfall on the shores of New Jersey in 2012. While hurricanes fairly often attack the southeastern coast of the US, Hurricane Sandy was very unusual in battering the central and northern parts of the East Coast. Why? In this section, we'll explore the rivers in the sky that Sandy drove onto such an unusual path and see one of the deadly phenomena that occurs when cold and warm air collide along these rivers—tornadoes.

Discovered in the 1920s by Wasaburo Ooishi, a Japanese meteorologist who was studying the atmosphere near Mount Fuji, jet streams were little known outside of Japan even a couple decades later, at the onset of World War II. The Japanese, however, were able to turn their knowledge of the jet streams to their advantage during the war by launching balloon attacks on the US, sending 9,000 "fire balloons" aloft to travel thousands of miles east.[22] Some 300 made it to US soil, and six people died when a family approached one and it exploded. (These were the only known deaths by enemy action on continental US soil during World War II.) As air flights increased during the war—in particular, flights between the US and the UK, where westerly tailwinds exceeding 100 mph made flights from west to east much faster than those in the opposite direction—knowledge of the behavior of jet streams became crucial.

Jet streams are giant rivers of air in the sky that mark bound-

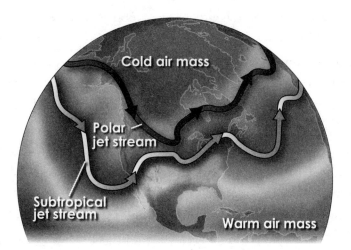

FIGURE 8.8 Schematic illustrating the world's jet streams. The deepest trough in the polar jet stream is an example of a negatively tilted trough discussed in the text. *From S. Marshak,* Earth: Portrait of a Planet *(New York: W. W. Norton, 2012).*

aries between cold and warm air. Although it is a big oversimplification, it is useful to think of the subtropical jet stream as lying roughly at the boundary between the Hadley and Ferrel cells at latitude 30°, and the polar jet stream as lying between the Ferrel and polar cells at 60°. Their positions, especially that of the polar jet, vary substantially over the year. In autumn, the polar jet moves from Canada to the south, bringing the cold air of winter into the US. In the spring, it retreats back north, allowing warm, tropical air to move northward into Canada for the summer.

Flowing at the top of the troposphere, the jets have variable elevations between 12,000 and 80,000 feet, with the subtropical jet being higher than the polar jet because the troposphere is higher in the tropics than in the polar regions. The jet streams can be several hundred miles wide and 1–2 miles deep, and they can flow at speeds of up to 400 mph. Jet stream winds generally flow from west to east, but they have a loopy structure and flow in various directions, even "backward," from east to west, in some segments. The loopiness, known as a Rossby wave, has a wavelength of about 1,800–2,400 miles and arises primarily because the Coriolis effect has different strengths at different latitudes. The jets can split apart, re-join, reverse, or simply stop.

In the Northern Hemisphere, segments of loops extending south are called "troughs," and segments extending north are "ridges." (You will hear these terms used frequently by meteorologists and weather reporters.) Without troughs and ridges, winds in the jet stream would move primarily from west to east, but the troughs and ridges cause the winds to circulate around low- or high-pressure cores. In the Northern Hemisphere, troughs host major low-pressure systems that have counterclockwise circulation, and ridges host high-pressure systems that have clockwise circulation. The direction of circulation is reversed in the Southern Hemisphere. Typically, a trough-ridge system might cover the whole US. Within it, smaller-scale circulation features ("cyclonic" features) can form at many scales—even down to individual tornadoes.

Hurricanes are powerful tropical storms that fall into the general category of "tropical cyclones," where the term "tropical" refers to the regions in which they form, and the term "cyclone"—originating from the Greek word *kykloun*—means "going around." *Kykloun* refers to the rotation of winds in a circle around a core that characterizes the storms. Hurricanes, cyclones, and typhoons are the same phenomena; the different names are used to describe these storms in the Atlantic, Indian, and Pacific Oceans, respectively.

Normally, an Atlantic hurricane starts in the tropics, where it is blown from east to west by the prevailing easterly trade winds. Then, as it moves north into the midlatitudes, it is blown in the opposite direction by the midlatitude westerlies (see Figure 7.10).

The westerlies push most hurricanes to the northeast, away from the East Coast of the US, where they sputter out at sea or attack eastern Canada. Such was the case in 2011 with Hurricane Irene, which, after soaking Quebec and the Maritime Provinces of Canada, died in the Labrador Sea.

Ridges and troughs in the jet stream migrate from west to east, but occasionally a high-pressure ridge just stops dead in its tracks. As a result, the trough behind it might bend backward on itself, rather like members of a marching band might fall backward if the front ranks unexpectedly stop. Such an event results in the formation of a so-called negatively tilted trough ("negatively tilted" refers to the fact that the axis of the trough points backward against the direction of motion; Figure 8.8). On the eastern side of such a trough, winds flow from southeast to northwest, so if a storm system happens to lie in the wrong place at the base of the trough, it can be "sucked" to the northwest by these winds.

Such was the case in late October 2012. A negatively tilted trough extended south all the way from Canada down into northern Florida, where it was in close proximity to the subtropical jet stream flowing over southern Florida. This configuration of the two jet streams is actually quite common and produces some of the strong storms that plague the Atlantic seaboard known as "nor'easters," such as the "Snowmageddon" that shut down Washington, DC, in February 2010,[23] or the potent winter storm of February 2013 that dumped up to 40 inches of snow in the Northeast.[24]

It is, however, highly unusual to have a hurricane sitting right in the middle of the two jet streams at the base of the trough. This was, unfortunately, the position of Hurricane Sandy during the last week in October 2012. The southeasterlies associated with the negatively tilted trough just sucked up Sandy, directing it to the northwest and the East Coast.

Just in case that position of Sandy in the middle of the two jet streams wasn't enough to seal its fate, the Atlantic Ocean was a veritable traffic jam of weather systems that fateful week, including a high-pressure/low-pressure couplet that had stalled over Greenland. Normally, the winter storm moving in behind the negatively tilted trough could move out to the northeast. However, the one associated with this particular negatively tilted trough was blocked by the couplet over Greenland. Sandy was, at the same time, being sucked up from the southeast. The winter storm and Sandy collided, wrapped around into each other, and merged into an energetic monster storm dubbed "Frankenstorm." Sandy took more than 85 lives, dumped a foot of rain in some places and 2 feet of snow in others, and moved up to 4 feet of sand from the beaches inland into coastal towns. It caused at least $30–$50 billion in losses.[25] Economically, Sandy was the second-most catastrophic disaster in US history after Hurricane Katrina, which cost at least $108 billion and took more than 1,300 lives.[26]

TORNADOES AND JOPLIN

The polar jet stream is not only instrumental in producing nor'easters on the Atlantic seaboard, but in the Midwest it produces thunderstorms and tornadoes by bringing cold Canadian air into contact with warm air from the Gulf of Mexico (Figure 8.9).

Tornadoes tend to form during the time of shifting of the jet stream—particularly in the spring when the northern air is still cold but the southern air is rapidly warming as the days become longer and the solar heat input increases as the sun rises in the sky. Thus, unfortunately for many midwestern cities, the months of April through June are prime time for severe storm activity. The Joplin tornado described at the beginning of this chapter was the second deadly tornado outbreak of 2011, occurring only three

FIGURE 8.9 Satellite view of the line of thunderstorms shortly before the tornadoes hit Joplin, Missouri, in May 2011. *Image from NOAA National Weather Service.*

weeks after a late-April outbreak that killed 327 people across six states—238 of those deaths in Alabama alone.

On the fateful day for Joplin, the four atmospheric ingredients that give rise to a perfect storm were in place: warm air, a mass of cold air, moisture, and winds.[27] It was hot and humid, and the sun had been out all day, building up the moisture content of the atmosphere. The masses of warm and cold air were present in a ridge that lay over the eastern US, and in a trough over the western states. The larger-scale weather pattern set by the jet stream provided an environment of high winds blowing in different directions that could spin up a thunderstorm into a huge rotating vortex known as a "supercell."

Low-level winds are dangerous by themselves, but they also provide another essential ingredient for tornado formation—a condition known as "low-level shearing." In this context, "shear-

FIGURE 8.10 The lower clouds in this photo are "wall clouds," which often indicate the presence and location of a rotating mesocyclone. Sometimes the wall clouds themselves are obviously rotating. *Photo by Brad Smull, NOAA Photo Library.*

ing" means winds blowing with different strength at different elevations. On the day of the Joplin tornadoes, the winds had a small velocity near the surface, but higher velocities at an elevation of a few thousand feet. This shearing produced a horizontal spinning vortex much like the rotors in the drowning machine and lee waves of mountains described earlier (see Figure 8.3).

That day also, there were regions where winds were rising—"updrafts" just like those you can produce if you turn on the venting hood over your stove. These updrafts draw air flowing along the surface and its vortices in and up (imagine that the horizontal vortex of a drowning machine is now sucked up vertically). Since the horizontal vortex is rotating, it imparts rotation to the updraft, forming a larger rotating zone known as a "mesocyclone," a vortex of air typically 2–10 miles in diameter (Figure 8.10).

(a)

(b)

FIGURE 8.11 Joplin, Missouri, with a southwestern branch of St. John's Regional Hospital in the center before (a) and after (b) the 2011 tornado. Distance across (a) is about 500 yards and (b) about 900 yards. *Images from Google Earth.*

The final ingredient in creating a setting for a tornado outbreak is a strong downdraft on the rear flank of the supercell. In ways not yet well understood by meteorologists, the presence of a strong downdraft here seems to concentrate the rotation within the mesocyclone, leading to the spawning of tornadoes.

The ingredients that lead to severe weather outbreaks are combined mathematically by meteorologists in a scale known as the

EHI (energy helicity index) scale. An EHI greater than 1 indicates supercell potential, an EHI up to 5 indicates the possibility of F2 and F3 tornadoes, and an EHI greater than 5 indicates the possibility of F4 and F5 tornadoes, the largest ones possible. The EHI for Joplin that day was 10.

The results were catastrophic (Figure 8.11).

REFLECTIONS:
TO WARN OR NOT TO WARN?

There are minutes to hours to even days of warning for some hazards, such as tornadoes, tsunamis, hurricanes, and some volcanic eruptions. For others, the time or place cannot be specified this precisely. Earthquakes, rogue waves, and, to a certain extent, landslides and other volcanic eruptions fall into this latter category. No single set of warning procedures applies to all disasters. Rather, warning systems need to be tailored to the individual hazard and population affected.

Our understanding of weather and meteorology has grown enormously over the past three decades, as has our understanding of the limits of prediction. These are topics in the next chapter. It remains, however, a major challenge for disaster planners to mesh science with human behavior. The experiences of seismologists in California and meteorologists have demonstrated that continual education, emergency preparedness, and drills work. But humans also tend to ignore warnings in the face of too much information and too many false alarms. I find it amusing that when the tornado sirens went off at work while I lived in Illinois, I, who am writing a book on disasters, and my colleagues in the geological community tend to run outside to see what's happening instead of following protocol and seeking shelter in the basement. The reason we feel justified in this apparently irrational action is that the warnings are generally issued on a countywide basis, and our county is so large that the statistical chance of the tornado being near us is small. Such is human nature.

The practice of tornado warnings evolved in spurts.[28] In 1887 the US Army Signal Corps banned official warnings for fear of causing panic and developing complacency among the populace, and only in 1934 was that ban partially lifted. However, many tornadoes occurred without warning because they could not be predicted. After the first successful prediction of a tornado at Tinker Air Force Base, Oklahoma, in 1948, the ban was completely lifted in 1950. During the 1950s and 1960s, warning systems were issued via commercial TV and radio, but for fear of confusion about military attacks on the US for which air raid sirens had been developed, the sirens that had been put in place in many communities during and after World War II were not used until about 1970. Now residents in tornado-prone areas can receive messages not only from sirens, TV, and radio but also through mobile, Internet, and GPS-based equipment.

Meteorologists saw the tornado outbreaks around Joplin coming days in advance, warnings of imminent danger were given more than twenty minutes ahead of time, and yet lives were lost. Interviews with survivors revealed that not everyone heard the sirens, people did not react until they actually could see the danger, and some did not take shelter because they had been desensitized to the warnings. Slightly more than half of those who died were in their residences. To address the issue of home safety, the building codes in Joplin now require fasteners called "hurricane ties" to attach houses to their foundation, although concrete basements that would provide strong protection are still not required in new home construction.

Whether for tornadoes, hurricanes, volcanic eruptions, or earthquakes, convincing people that warnings are not released casually, that they need to prepare in advance, and that they need to have an action plan and implement it when a warning is issued remains a challenge. The warnings won't always be correct, there will be false alarms, and there will always be second-guessing, but existing warning systems and preparedness measurements do reduce casualties and economic costs. With good research and education of the public, these measures and their results can only get better.

Chapter 9

WATER, WATER EVERYWHERE . . . OR NOT A DROP TO DRINK

A PLAGUE OF SNAKES

At the end of 2010 and the beginning of 2011, record-breaking floods devastated Queensland, a state in northeastern Australia. Warnings went out and emergency plans were activated. In contrast to flood preparations where I lived in the Midwest, such preparations in Australia include ordering emergency supplies of snakebite antivenom! Queensland has forty different species of snakes, most of which are venomous. Many of these snakes are not small—the eastern brown snake is usually 4–5 feet long, and others are 6–8 feet. People stranded in boats couldn't take shelter in the trees because the snakes, fleeing from flooded nesting places, found the trees first. Snakes can swim, and they are good at finding trees to climb and houses to enter. They like attics. And most of these snakes were "cranky," a term used by the Aussie press at the time, because they had been disturbed during their mating season.

Of a woman who was temporarily living on a boat because of the floods: "She could feel the tongue flicking on to her face to test how far away it was, ready to bite, and then it jumped into her lap."[1] Australians are, if nothing else, action oriented. The woman's husband flicked the snake into the water with a stick. The report was that their screaming and panic nearly capsized the boat, but my vision of the Aussies is that they actually took this all in stride; after all, what's just one more snake when there are crocodiles that are well camouflaged in the floodwaters because they look just like floating debris, yucky cane toads everywhere, and sand flies and mosquitoes providing discomfort on another level?

Snakes aside, while in Australia during July of 2011, I discovered yet another consequence of these floods: bananas were about Aus$15 per kilogram (US$6 per pound), or about US$2.50 each (Figure 9.1). Although part of the story about these high prices probably involves powerful political influences, Australians do try to minimize their agricultural imports for fear of importing pests into their ecologically fragile continent. This practice makes prices fairly high at all times because there's not much competition from foreign goods, but it made them extraordinary for bananas that summer because the local banana crop had been devastated by the floods.

Droughts are the mirror images of floods. In 2012, Illinois was stricken by a major drought that affected the farmlands, but no one was short of drinking water. However, I've had enough experience doing fieldwork in the deserts of the western US to know vividly how the absence of water can affect your life. After my first year as a graduate student at Caltech, I wanted to get some field experience doing classic geological mapping. My adviser, not daunted by the fact that I had never camped before, took me to the far southern end of what is now Canyonlands National Park. This is a beautiful, but barren and dry, area, and in the

FIGURE 9.1 Bananas for sale in Melbourne, Australia, in 2011. *Photo by Gerard Lopez.*

summer the daytime temperatures reach 110°F by ten o'clock in the morning. I lived there alone for four weeks.

My campsite was two hours from the nearest water by four-wheel drive and four hours from civilization in Moab, Utah, which in 1965 was a sleepy, dusty frontier town. Since I was facing a month alone in this remote camp, I adopted the only stray dog in Moab—a border collie puppy. Thus, I had to have water not only for myself, but for a fragile puppy. Food provisions weren't much of a problem; canned foods do fine (except that the chipmunks ate all the labels off the cans during the first week, so it was a bit of an inadvertent potluck all the rest of the summer). I learned to take a fairly complete bath with a cup of water (possible only in a desert environment!), to keep my dog and myself hydrated, and to treasure every sip of water. I can only imagine how difficult life must be in the drought-stricken regions of Africa where, day after day, the quest for water dominates the lives of people, particularly women and children, who often must walk long distances every day to fetch the water.

Utah has an arid climate, and with the exception of some occasional thunderstorm weather, it was consistently hot and dry the summer I was there. The term "weather" applies to conditions in the atmosphere over relatively short periods of time up to months, whereas the term "climate" refers to long-term averages of weather. The thunderstorms were weather; the arid conditions were climate. The term "climate variability" refers to short-term variations in the climate, with timescales of a year or a few years. Both weather and climate represent changes of state in the atmosphere, but at different timescales. What's going on with the weather? What controls our weather? Why is it so difficult to predict? Are recent floods and droughts around the world best described as weather, climate variability, or an indication of climate change? The answers to these questions are topics of intense debate in the scientific community. The economic and political implications are enormous.

We'll spend most of our field trip in this chapter trying to understand conditions in the Pacific Ocean off the western shore of South America. I will not enter the climate-change debate; rather, I hope to provide you a basis for understanding just how complex the underlying science is and, thus, why scientists can have divergent opinions when it comes to projecting the future of weather patterns and climate.

2011: BILLIONS AND BILLIONS

Large storms and droughts have been such a part of the relation of humans to nature that stories of them pervade our oral and written traditions. The year 2011 was extraordinary. Australia and the US were not the only places to suffer record weather-related disasters that year. Massive flooding hit Brazil with rain and mudslides in January. Rain, landslides, and mudflows devastated parts of China in August. And prolonged flooding displaced nearly 13 million people in Thailand.

According to NOAA, in the US alone there were fourteen billion-dollar disasters in 2011, and at least eleven in 2012—all weather related. Of the 2011 events, six were the massive tornado outbreaks in the central and southeastern part of the US covered in Chapter 8. The rest involved weather in one way or another: floods, a blizzard (which can be thought of as a snowy variation of a flood), Hurricane Irene, the Southwest drought and related wildfires in Texas. Floods and droughts were both occurring in these years.

Let's look at the events of 2011 a little more closely. In late January the Midwest was hit by a historic blizzard, now dubbed the "Groundhog Day Blizzard." By this time, midwesterners had already had two long months of snow and were warned that another 2 feet might be on the way. On January 31 airlines began to allow travelers in the potentially affected areas to change flights for free during the next three days—an economically expensive decision for them and a sure sign that something big was on the way. An editorial meteorologist said, "When everything is said and done, the storm may well impact a third of the population of the United States—approximately 100 million people."[2] Officials in Chicago urged people to stock up on food and medicine, and warned that the storm might cause 25-foot waves on Lake Michigan. Freezing rain was predicted for other parts of the Midwest. The National Weather Service called the storm "dangerous, multifaceted and potentially life-threatening."[3]

Then, just as winter was receding back into the north, spring came with the tornadoes, followed by billion-dollar floods in

North Dakota. The Red River that forms the border between North Dakota and Minnesota is one of a few rivers in the US that flow northward, continuing into Manitoba, Canada, where it empties into Lake Winnipeg and eventually into Hudson Bay and the Arctic Ocean. The Red River has a dynamic very different from that of rivers that flow southward into warmer waters. Because it flows into colder climes rather than warmer ones, the river encounters ice jams and frozen lakes that dam its path to the ocean along the way. Towns along the Souris River, a tributary of the Red River, were hit especially hard. The scene for this flooding had been set the previous year when summer flooding saturated the ground. Then, a big snowpack during the 2010–11 winter was saturated by heavy rains and thunderstorms in local areas. Some areas received their normal annual rainfall in only two months,[4] and flooding ensued.

Allowing FEMA workers little time to rest, Hurricane Irene swept up the East Coast later in the summer, causing extensive flood and wind damage. Its projected path put more than 65 million people from the Carolinas into Atlantic Canada at risk. The hurricane made landfall in North Carolina, and with the major metropolitan areas of Washington, Philadelphia, and New York City in its projected path, preparations were made for flooding of low-lying areas all the way up the East Coast. In many areas the ground was already saturated from a record wettest summer, and serious flooding was anticipated. Air, car, and train transportation were disrupted up and down the coast. Nearly 6 million people lost electricity, and hundreds of thousands were evacuated. Philadelphia experienced record levels of flooding, southern Vermont was hit hard, but damage in New York City was less than feared.

During the same time that the eastern half of the country was experiencing all too much rain and flooding,[5] Texas, Oklahoma,

Kansas, and the Southwest were experiencing record droughts and wildfires. Starting in the autumn of 2010, ten months of drought dried up lakes, parched the grasslands, and allowed wildfires to burn across the region, especially Arizona (which had its largest wildfire season in history) and Texas. Oklahoma had the warmest May–July period in its history, while eighteen other states had a "top ten" warmest three months during that year. In Dallas, temperatures exceeded 100°F on thirty of the thirty-one days in July.

One way that climatologists measure drought is by comparing annual rainfall to a long-term norm. A severe drought in Texas early in the twentieth century was caused by a deficit of 10–12 inches of rain compared to the long-term norm, and one in the 1950s was brought on by a 6- to 8-inch deficit. In 2011, the deficit was 13 inches, making the resulting drought the most severe recorded since record keeping had begun in 1896.[6] The 2011 drought also covered a huge area compared to previous droughts in Texas (Figure 9.2). By August of 2012, 62 percent of the contiguous US was declared to be under "moderate to exceptional drought."[7]

MIRROR IMAGES: DROUGHTS AND FLOODS

To understand how there can be floods in some places while there are droughts in others not very far away, we need to examine both the local and global pictures. There are many ways to look landscapes—for beauty, for city planning, for recreation. Another way is to look where water flows. Rain falls onto the land—sometimes over large areas as in hurricane-scale weather events, sometimes over small areas at the scale of individual thunderstorm cells.

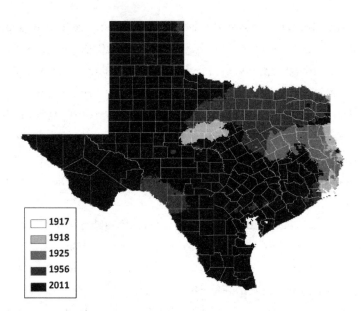

FIGURE 9.2 Areas of Texas that have seen droughts. Shading represents a different year, as indicated in the lower left. Note the huge area of the 2011 droughts compared to other years. *Map provided by Brent McRoberts, Texas A&M University.*

Water flows downhill toward low points, forming small rivulets, creeks, streams, and ultimately larger and larger rivers that end up in lakes or oceans.[8] An area where all of the water flows to the same place is called a "watershed." Watersheds are separated by high regions called "divides"—the Continental Divide of North America being a prominent and large-scale example. Water that lands west of the Continental Divide ends up in the Pacific Ocean or Bering Sea, and water that lands east of it ends up in the Gulf of Mexico or Atlantic and Arctic Oceans.[9]

A flood occurs when a storm delivers more water to a watershed than can be carried in the normal channels of its creeks and rivers. Because we have altered natural drainages with massive engineering manipulations of our rivers and streams, "normal" is a relative term. For practical purposes, it means whatever conditions the ecosystems have gotten used to over the past decades. Such a definition allows for "normal" to be applied to both the hot, wet Amazon basin and the hot, dry deserts of the southwestern US or Africa.

Sometimes a river channel is clearly defined—for example, where a river flows in a deep incision in a landscape of solid rock, like the Colorado River flowing through the Grand Canyon. Other times it is less well defined, as where the landscape is formed of soft soil and the course of a stream or river varies as water erodes some parts of the system and deposits sediment in other parts; a good example is the Mississippi River over most of its path to the Gulf of Mexico. A flood occurs when water flows out of the channel onto normally dry land. The concept of "normally dry land" is, again, vague, but it is loosely defined as land that experiences floods only every few decades to hundred years. This land is referred to as the "floodplain." Civilization started in the floodplains of river valleys, and flooding was a common occurrence. Over much of human history, dam building and modification of river channels has been aimed at reducing the exposure of humans to floods, yet even today a large fraction of the world population lives in floodplain environments.

Watersheds respond differently to storms of different sizes. If a storm delivers water over an entire watershed, all rivers in that watershed may flood. But if a storm is small compared to a watershed, the individual rivers and streams may respond in dramatically different ways. Some rivers may show no flooding; others may flood moderately if some of the streams feeding into them are flooding. Yet others may show extreme flooding if they suffer a direct hit from a storm. Some rivers rise slowly, inexorably but gradually climbing their banks over hours or days, depending on the size and history of the storm. Others seem to rise nearly instantaneously, often with a wall of water suddenly appearing. These are "flash floods."

I had a vivid illustration of the variability of rainstorms once when I was stranded by a classic flash flood during my stay at Canyonlands. Starved for company, I had driven a few miles across a few arroyos (a western word for a creek bed) in hot, sunny weather to join a visitor for dinner. We sat watching the sunset and a beautiful southwestern thundercloud and storm in the distance. About an hour later, in the dark, we heard an ominous noise—water rushing down the small arroyo not too far away, an arroyo that separated me from my campsite. I spent the night on the wrong side of the arroyo, a victim of a storm that hadn't even gotten me wet! This is an example of wet weather in a dry climate.

Droughts can be thought of as mirror images of floods. They occur when a region has less water than normal. Both the Amazon basin and the Mojave Desert can suffer from droughts when precipitation is reduced below the value of normal for any significant period of time. Soils dry up, winds can blow the soil away, plants and animals dependent on the moisture die, animals migrate away in an attempt to find water, and humans suffer from lost agriculture. Water for industrial users can become scarce, and if reservoirs behind hydroelectric dams dry up, electric power generation is reduced or halted. These conditions can lead to social unrest and human migrations, including warring over water and food.

In the short years of the twenty-first century, there have already been notable droughts. The prolonged drought in the Horn of Africa and to its northwest significantly contributed to the Darfur conflict in Sudan and Chad as the Arab nomads went farther south into lands occupied by non-Arab stationary farmers. In 2005, parts of the Amazon experienced the worst drought in a century, with another one even more severe and covering four times the area occurring in 2010. This ecological system is so fragile that scientists predicted[10] in 2004, before these two droughts, that the rain forest could be stressed to a tipping point where it might turn into a grassland. Such a transition would have enormous consequences to the CO_2 budget in the atmosphere. It is unknown how long the American Southwest drought centered in Texas will go on, but one wonders if the future might be similar to that which the environmentalist author Tim Flannery[11] has predicted for Perth, Western Australia. Flannery worries that Perth could become the world's first "ghost metropolis," an abandoned city because of the loss of water resources. Is this a scenario that Phoenix, Las Vegas, or Albuquerque should be contemplating?

In 2011, meteorologists anticipated that the Texas drought might continue because another La Niña was developing in the Pacific—conditions historically associated with Texas droughts. In fact, the second La Niña extended well into 2012. What is La Niña? How does it cause droughts in Texas? Could it also cause flooding on the East Coast? What about its partner, El Niño? What can we say about these questions? And what can't we say? It is impossible to review all of atmospheric dynamics here, but I will focus on two of the most important: El Niño and La Niña.

CHILDREN OF THE TROPICS:[12]
EL NIÑO AND LA NIÑA

Perhaps no people know more about surviving the flood-drought cycles than the ancient aboriginal people of Australia, who have survived in their desert homeland for more than 60,000 years by knowing every element of their land, including the sites of precious "permanent waters."[13] There are no permanent rivers in central Australia, but sometimes rivers will flow for a day or, occasionally, for a few months. These are the permanent waters to the aborigines. Although recently coming to live more and more in metropolitan areas, they were historically a nomadic people, venturing quickly out into the arid parts of the land when rains came, and retreating to the different permanent waters during droughts.

In the late eighteenth century, British colonists came to Australia and discovered the droughts—a particularly severe one occurring a century later, in 1877. We now know that this drought caused the deaths of over 9 million people in China and 8 million in India.[14] Henry Blanford, head of the India Meteorological Department, noticed that atmospheric pressures were higher than usual over India at this time, and he sent word around the British Empire asking whether others were also experiencing droughts. Charles Todd, the South Australian government meteorologist, responded that there was a drought in Australia. During the next drought, in 1888, Todd noted that there was also one in India. These early studies of "teleconnections," as well as a number of others in the early twentieth century, led to the discovery of two ocean phenomena known as La Niña and El Niño, and the atmospheric phenomenon known as the Southern Oscillation. The ocean and atmosphere changes are linked, and the combined changes are referred to as the El Niño/La Niña Southern Oscillation, or ENSO. Because the Southern Hemisphere is dominated by oceans, the effect of ENSO is strongest there, particularly in the South Pacific and Australia.

Under "normal" conditions (Figure 9.3a)—that is, in years with no El Niño or La Niña—the trade winds around the equator blow from east to west, pushing surface waters of the Pacific Ocean to the west toward Indonesia. As the water travels westward, it warms, typically down to depths of roughly 500 feet.

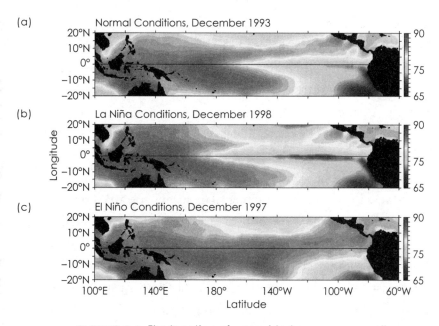

FIGURE 9.3 The location of warm (darker gray, generally toward the left) and cool (lighter gray and darker black, generally toward the right) waters during normal conditions (a), La Niña conditions (b), and El Niño conditions (c). NOAA Pacific Marine Environmental Laboratory (PMEL).

One way to think about this is to imagine that there's a huge, fairly deep bathtub filled with water stretching along the equator from South America all the way across the Pacific, across the Indo-Pacific islands and Australia, into the Indian Ocean. Imagine that your head is in this tub at South America, your belly button is at the mid-Pacific around Tahiti, your knees are somewhat north of Australia, your feet are around India, and the equator runs down the middle of your body. Imagine also that this water is like a swimming pond outdoors—warm on top and cooler underneath—and that you turn on a fan blowing from your head to your toes across the bathtub to represent the trade winds blowing from South and Central America toward Indonesia and India.

Warm water moves toward your toes, and cooler water down below moves toward your head to replace the water gone west (this is conservation of mass). This upwelling water carries nutrients that supply the rich fisheries off the west coast of South and Central America.

When the atmospheric pressure is very high in the eastern Pacific compared to the Indian Ocean, a variation of normal called "La Niña" (meaning "girl child") occurs (Figure 9.3*b*).[15] It pushes the warm water in the bathtub farther than normal toward your feet. In the real world, the warm pool of water is piled up near Indonesia, providing moisture to fuel heavy precipitation in the western Pacific. La Niña conditions yield heavy monsoons in India and Southeast Asia and wet weather in northeastern Australia. The 2010–11 La Niña was one of the strongest ever recorded, causing the failure of the banana crops in Queensland.

Normal and La Niña years occur when there is low air pressure in the western Pacific. In some years, however, for reasons still mystifying scientists, the surface pressure in the western Pacific and over the Indian Ocean, Indonesia, and Australia rises, while the air pressure in the eastern Pacific decreases (Figure 9.3*c*). Trade winds weaken, or even reverse directions and flow from west to east. When the trade winds are weak, warm surface water can flow back toward the east, where it lies off the coast of South America and provides moisture for wet years there. The fishery collapses because the colder waters no longer well up toward the surface. Since this happens around Christmastime, the fishermen in South America who recognized this phenomenon gave it the name "El Niño," which means "the little boy" or "Christ child" in Spanish.[16]

Even though the driving forces for El Niño and La Niña occur in the Southern Hemisphere, their effects are global. During El Niño years (Figure 9.4*a*), northern regions in the US and Can-

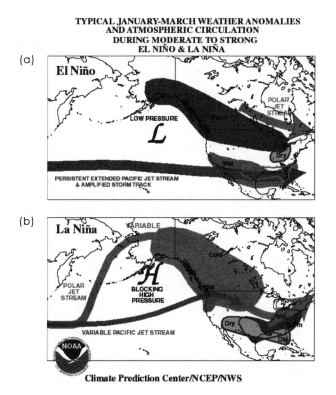

FIGURE 9.4 The predicted effect of El Niño (a) and La Niña (b) on North America, according to one model. NOAA.

ada along their border are warmer and have less-than-normal amounts of snow. When Vancouver, Canada, hosted the 2010 Winter Olympics, an El Niño season potentially threatened the games because of the lack of snowfall. Wetter winters occur in El Niño years in the American Southwest, including central and southern California. In contrast, during La Niña years (Figure 9.4b), the northwestern part of the US and western Canada are cold and the southern US is mostly dry.

The ocean-atmosphere system is more complicated in the Northern Hemisphere than the Southern—particularly in the North

Atlantic, because there is a relative of ENSO called the North Atlantic Oscillation (NAO). The weather "oscillates" between two states—a phenomenon that significantly influences the weather in Europe and, to a lesser extent, in the eastern US and Canada. In fact, the traffic jam of weather systems over Greenland that prevented Hurricane Sandy from moving off to the northeast (discussed in the previous chapter) was due to the NAO.

The weather in Europe is generally driven by westerly winds bringing moist air from the Atlantic into Europe. To understand the North Atlantic Oscillation, use the bathtub model already described, but this time position your head around Greenland or southern Iceland and your feet down in the subtropics, by Portugal, Gibraltar, or the Azores. In both states of the NAO, the pressure is low over your head (southwest Iceland) and high over your feet (the Azores). The cycles of this oscillation don't have names, but are just referred to as "positive" or "negative." The Azores high pressure moves back and forth in an east–west direction, like a pendulum anchored in Iceland sweeping back and forth across the Atlantic. This motion has a significant effect on the weather in Europe and the eastern part of the US and Canada because it influences the position of the jet stream.

When there is a strong difference in pressure between the Icelandic low and the Azores high, the westerly winds are strong and the summers and winters in Europe, as well as the winters on the East Coast of the US, are relatively mild because the jet stream, which would bring in cold air, is kept back to the north. This is the positive phase of the oscillation. When conditions reverse into the negative phase, there are fewer and weaker winter storms in the Atlantic, but they cross farther south. Cold air flows into northern Europe, the East Coast of the US is cold and snowy, and citrus crops in Florida suffer.

Unfortunately for forecasters, weather is notoriously complex because there are many more phenomena than those discussed here, and they interact in complicated ways. For example, during the El Niño of 2006–07, when southern California should have been wetter than usual, it ended up being the driest on record since 1877. Many forecast maps illustrating possible El Niño and La Niña conditions for your region exist on the Internet, such as those that I've selected for Figure 9.4, but these are, at best, average estimated conditions. Any one season or year can be dramatically different because local variabilities and teleconnections are still not fully accounted for in even the most sophisticated computer projections.

WHIRLING AROUND

An inevitable and intriguing question is, Do cycles like ENSO, NAO, and others affect the number or intensity of hurricanes and cyclones? The answer is a qualified yes, but to understand how much and why, we need to know some of the basic dynamics of these big storms.[17]

In each of the major ocean basins around the world, the rated severity of tropical cyclones, and the terms that describe them, are different depending on the strength of the winds. For example, a tropical storm in the Pacific becomes a "typhoon" (and one in the Atlantic becomes a "hurricane") when the winds reach 74 miles an hour for at least ten minutes.

Climatologists struggle to project the influence of climate change on hurricanes because there are so many factors that affect the dynamics. Meteorologists have difficulty projecting the exact path and intensity of a hurricane for the same reasons. The factors that contribute to hurricane formation and strength are several, are complicated, and often involve quantities, such as ocean temperature profiles, that are ill known. But the major factors are the position and abundance of warm water, the structure of the atmosphere during hurricane season, and proximity to the equator.

Hurricanes are fueled by warm water. Water temperatures above 80°F are generally required to supply a hurricane. Over a body of warm water, the atmosphere can become warm and humid, setting up the stage for atmospheric convection to take the moisture to great heights. This is the basic setting for thunderstorms, a component of bigger storms. Like most bodies of water on Earth's surface, the ocean cools downward. (Ocean swimmers treading water know that their arms and head can be in warm water, but their feet in much cooler water.) A shallow, warm surface layer like this isn't enough to fuel a hurricane. For a warm water body to affect a storm system, it needs to extend downward more than 500 feet. If not, budding storms have the capability of mixing the shallow warm water with deeper, cold waters and cooling the average temperature to below the 80°F threshold.

Not all rising air produces thunderstorms, let alone hurricanes. The atmospheric conditions have to be such that the water in the rising warm air condenses, as it does when steam exits the nozzle of a teakettle. This process, as discussed in Chapter 8, releases latent heat. This heat then warms more air that, in turn, rises, powering yet more convection.

Convection can be disrupted, however, by strong horizontal winds in the atmosphere—the phenomenon known as "wind shear." This is not quite the same wind shear that terrifies airline pilots and passengers; that wind shear is in a vertical direction and endangers planes by throwing them downward toward the ground. The wind shear that disrupts hurricanes is horizontal; it's the kind of shear that blows the smoke from a chimney sideways and prevents it from rising high above the chimney. And it is here, in the wind shear story, that there's a connection back to El Niño.

Several processes result in the formation of hurricanes,[18] but the role of the Sahara in western Africa is by far the most dramatic. The conditions in the hot, dry Sahara and the warm, moist countries around the Gulf of Guinea play a critical role in the formation of hurricanes. It is here that most (85 percent) of the hurricanes in the Atlantic and, surprisingly, a significant number of cyclones in the eastern Pacific start forming after crossing the Atlantic and Caribbean, as well as Central America.[19] Every few days an atmospheric disturbance that could become a hurricane blows west from Africa out into the Atlantic. These disturbances pick up energy as they hit the warm waters of the Caribbean and Gulf of Mexico.

If there are winds at high elevations coming from the Pacific across Mexico into the Gulf of Mexico and the Caribbean—as, for example, in El Niño years—a developing storm, moving west, becomes smeared out by these high-level winds blowing toward the east. The energy of the storm spreads out over a large area, a process that prevents it from developing into a major hurricane.

Hurricanes develop out of weather systems that are already disturbed, the rather common types of storms that develop on the planet from air flowing from high pressure toward lower pressure. But, as we saw in Chapter 8, air doesn't flow in a straight line from high to low pressure. The Coriolis effect deflects the winds and, when conditions are right, spins them up into the vicious cyclonic frenzy needed for hurricane formation. There is no deflection of the winds by the Coriolis effect right at the equator. But about 5° of latitude away from the equator, this effect becomes strong enough to allow circulation to develop around low-pressure zones.

The evolution of a tropical storm depends on all of these factors.

If an atmospheric disturbance is too close to the equator, no Coriolis effect. If it is too far from the equator, no warm water. If there is too much horizontal wind shear in the atmosphere, the tops just blow off of budding storms. If there is too little humidity in the atmosphere, there's not enough moisture to release latent heat.

Hurricanes in the Atlantic are more common in La Niña years, like 2011, than in normal or El Niño years. A look at the life cycle of Hurricane Irene[20] reveals how these monster storms develop, and a look at Hurricane Sandy's crazy path shows why they are such a headache for weather forecasters and emergency planners.[21]

Like most Atlantic hurricanes, Irene was born off the big bulge on the northwestern coast of Africa, the home of the Sahara desert. South of the Sahara lie the hot and humid forests of Guinea, Ghana, Burkina Faso, Niger, and Nigeria, and the ocean along the Gulf of Guinea coast. Separating these two different climate regions in the atmosphere is the African easterly jet stream. Irene arose from a tropical wave that developed in this jet around August 15, 2011. At this stage, winds in the future Irene were disorganized, blowing in many different directions with velocities of less than 25 mph. Four days later, the winds began increasing and developing a circulating motion; thunderstorm activity within the system increased. In less than a week, Irene rose to the category of a tropical storm with sustained winds of 39 mph. On August 22, just after making landfall in Puerto Rico with winds of 70 mph, the storm was upgraded to a category 1 hurricane.

Tropical hurricanes are characterized by strong vertical temperature gradients between the ocean surface and the upper atmosphere, and they are fueled by the vertical convection of warm water. If Irene had stayed over land after making landfall in Puerto Rico, it would have weakened because the supply of warm water would have been cut off. However, Irene moved back

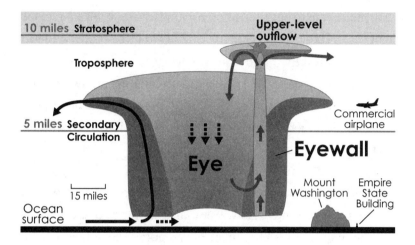

FIGURE 9.5 Structure of a hurricane. *NASA, Goddard Space Flight Center.*

out to sea, intensifying rapidly as she traversed very warm waters near the Bahamas, becoming a category 3 hurricane.

One of the most prominent features of hurricanes is their central "eye," a clear region of low pressure in which warm winds flow gently downward, giving these storms the name "warm-core storms" (Figure 9.5). Surrounding the eye is the "eyewall," a region of very strong upward winds. Moist air from the warm ocean surface spirals inward toward the eyewall, where it ascends. During ascent, condensation of the moisture provides heavy rain and releases latent heat that fuels the convection. At the top of the eyewall, air flows both outward, leaving the hurricane (for example, as illustrated in Figure 9.5 at the "hot tower"), and inward, where some is recirculated slowly down the eye.

Most major storms with winds over about 115 mph develop a second eyewall, so the original eye becomes an eye within an eye. Irene was no exception. In a battle of giants, the outer eyewall moved into, and choked off, the inner one. This process is called "eyewall replacement." Hurricanes weaken during eyewall replacement but can reintensify after the eyewall replacement cycle. Irene never really recovered from the battle, and as it approached the Outer Banks of North Carolina, was downgraded to a category 1 storm. Nevertheless, it remained a powerful and very large storm, causing over $15 billion in damage due to rain and flash flooding before it eventually exited into the Labrador Sea two weeks after it had left Africa.

Only a year later, Hurricane Sandy struck with very different, and much more damaging, consequences. What was so different about Sandy? As we saw in Chapter 8, one difference was the inauspicious alignment of the polar and subtropical jet streams, coupled with a traffic jam of weather systems in the North Atlantic caused by the North Atlantic Oscillation. These meteorological features put Sandy on a collision course with a much larger swath of major infrastructure along the eastern seaboard than was exposed to Irene.

Second, the collision of Sandy with the winter storm moving across the US behind the trough in the jet stream spun it up into a monster storm with high winds spanning 1,000 miles in diameter. As Sandy interacted with the cold air brought south in the negatively tilted trough of the polar stream, it lost its tropical warm core. Remember that, in one sense, the "purpose" of a hurricane is to transport heat from warm to cold regions to eliminate strong temperature gradients. The loss of the warm core should have killed Sandy because it diminished the vertical temperature gradient that fueled the storm. But Sandy latched on to a new source of energy: the horizontal temperature gradient between the warm subtropical air and the cold polar air in the trough. This is the driver of midlatitude, or "extratropical," storms. Instead of dying out as it lost its tropical-storm characteristics, Sandy became invigorated and turned into a massive storm with a "cool core," engulfing the cold polar air into its spiral arms and enlarging further, into a Frankenstorm.

A third reason for the enormous damage done by Sandy was its close approach to the coast and the fact that it made landfall at high tide. Winds in a hurricane push a bulge of water toward the shore—the "surge." When this surge is superposed on top of a high tide, floodwaters inland can be several feet higher than if the same event occurred at low tide. The incoming waters carried sand and debris from protective barrier islands and beaches to points far inland. The destruction of the islands and beaches presents a major design and remediation challenge as residents of the East Coast recover and plan to reduce the impact of hurricanes in the future.

WHAT CAN WE SAY, WHAT CAN'T WE SAY, AND WHY?

In 1967, mathematician and meteorologist Edward Lorenz wrote a monograph in which he laid out a set of equations that, to the best knowledge of the time, described the behavior of the atmosphere.[22] He opened a chapter entitled "The Problem" with the following quote by George Hadley, of the Hadley cells discussed in Chapter 7:

> I think the causes of the General Trade-Winds have not been fully explained by any of those who have wrote on that Subject.

Even now, when so much more is known about conditions in the atmosphere, and enormous supercomputers grind away at solving the equations, much remains uncertain about how the atmosphere operates and how small-scale phenomena, such as weather, relate to large-scale, global conditions.

Lorenz's challenge was to write down the equations and then simplify them enough to solve them with the rather limited, by modern standards, computers available at the time. The atmosphere itself is complex (especially its vertical temperature distribution), and the processes operating on it, and in it, are also complex. The sun pours radiation into the atmosphere with different intensities depending on latitude and time of day. This radiation is absorbed, emitted, scattered, and reflected. Eddies of many different sizes transfer energy like eddies do in a pot of boiling soup. It is impossible to capture all of these processes exactly in the equations.

Lorenz divided the physical laws that he needed to describe the atmosphere into two groups. The first are the basic laws that we discussed in Chapter 2: conservation of mass, momentum, and energy, as well as the equation of state for the gas of the atmosphere. The second group consists of the laws that describe the forces on the atmosphere and heating or cooling of it, the evaporation and condensation of water, and the conversion of cloud droplets into raindrops or snow crystals.

The purpose of the equations that Lorenz was attempting to solve was to take a given condition of the atmosphere—for example, today—and predict its future state. Lorenz found that his equations were extremely sensitive to small variations in the atmosphere's starting conditions—for a hypothetical example, whether a given place was at a temperature of 60°F or 61°F. The solution to the equations two weeks ahead, for instance, might be dramatically different for these two cases. This dilemma obviously caused great problems for weather forecasters. Although much progress has been made, the ability to forecast ahead is still limited. The work of Lorenz laid the foundation for a powerful new science known broadly as "chaos science."

Because of these complexities, scientists have typically said that they can make general statements like "hurricanes in the Atlantic are more frequent in La Niña years," but not specific statements like "Hurricane Irene was caused by a La Niña." For similar reasons, it has not yet been possible to definitely attribute specific weather events to climate change rather than to climate variability. Most scientists now have hopes that making such connections will become possible, but that ability does not yet exist at the time of publication. This challenge lies in the domain of a new and controversial field called "attribution science."[23]

The challenge is difficult because many factors influence a specific weather event. The Russian heat wave of 2010 is an example. One analysis of this event,[24] for example, concluded that the heat wave was a result of natural climate variability, not climate change. In another analysis,[25] scientists attempted to decide whether six extreme events of 2011 could be attributed to climate change. For the Thailand floods that had submerged parts of Bangkok under nearly 10 feet of water for two months, the verdict was that the level of rainfall was fairly typical for a La Niña year, but that changing riverbed conditions in the Chao Phraya River and land use practices had made the area more prone to flooding that particular year. There was no demonstrable role of climate change.[26] On the other hand, it appears that climate-change warming of the Indo-Pacific Warm Pool is contributing to more frequent East African droughts, such as those affecting Somalia and Ethiopia.[27] And for the Texas drought of 2011, the verdict was that extreme heat events were roughly twenty times more likely in 2008 and 2011[28] than in comparable La Niña years of the 1960s.[29]

REFLECTIONS:
THE PRECAUTIONARY PRINCIPLE

Storms and floods account for nearly 70 percent of the world's disasters.[30] In addition to drowning, people die of diseases, malnutrition, and hunger because droughts and floods take a major toll on agriculture and food production. While high banana prices (Figure 9.1) may have been a minor annoyance to Australians, in a world with an ever-increasing population, food security is a major issue of the next decades. Some believe that the food shortages caused by droughts were a catalyst for the revolutions of the "Arab Spring" of 2011. It is clear that future food price increases will have an effect on the poorest people on the planet—those who live on $2 a day or less.[31]

Determination of whether human-induced climate change versus natural climate variability is involved is of vital importance because it determines whether we proceed inexorably on a path of more droughts and floods or might return to a previous state. Paul Krugman, winner of the 2008 Nobel Prize in Economic Sciences and *New York Times* columnist, pointed out that in 2011 we were in the midst of the second global food crisis in three years.[32] With inflation in the US economy low by historical standards, those in the US have not felt this food crisis as much as the world's poor, who spend as much, or even most, of their income on basic foods. Dismissing US monetary policy and speculators as the cause, Krugman highlights the extent that severe weather in recent years, particularly in the summer of 2011, has disrupted agricultural production. One-fifth of the world's land area experienced high-temperature records. Krugman concludes, "The evidence does, in fact, suggest that what we're getting now is a first taste of the disruption, economic and political, that we'll face in a warming world."[33] Statements such as these illustrate why it is essential that we determine how much our collective behavior versus other nonhuman effects contribute to future temperature changes on this planet that is our home.

Increasingly, governments and organizations have been recognizing and adopting a principle known as the "precautionary principle" to guide their actions. This principle states that if an action or policy decision is suspected to cause harm to the public or the environment, the burden of proof that it is *not* harmful falls on those taking that action, even in the absence of scientific consensus that the action is harmful. This principle is written into the laws of the European Union, was part of the 1987 Montreal Protocol on protection of the ozone layer, and constituted Principle 15 at the 1992 Rio Conference ("Earth Summit"). Although the precautionary principle is not commonly referred to in the US, the city of San Francisco adopted a requirement that city government base its purchase policy for everything from cleaning chemicals to computers on this principle. The degree to which nations adhere to the precautionary principle underlies some of the fundamental disagreements that have been making a global climate-change agreement so difficult. Explicit examination of these differences, and their resolution, might help break this roadblock to arriving at wise decisions about our use of resources that affect climate, floods, and droughts.

Chapter 10

EARTH
AND US

L'AQUILA: SCIENTISTS ON TRIAL

For centuries, residents of the medieval city L'Aquila had evacuated from their houses to a nearby piazza (or, more recently, to
their cars) to wait out sequences of earthquakes in accordance
with wisdom passed down in families living in the earthquake-
prone heart of Italy.[1] Earthquakes are known to have happened
here as far back as 1315, one in 1461 causing massive regional
destruction, one in 1703 killing about 5,000 people, and another
in 1786 that killed more than 6,000 people. One of the deadliest European earthquakes of the twentieth century killed 30,000
people just 25 miles south of L'Aquila in 1915. Starting in the
summer of 2008, minor to moderate seismic activity (magnitude
less than 4.5) rattled the ground, and earthquake lights in various
forms were spotted in and around the city. A resident who was
measuring radon gas levels was making unofficial forecasts of an
earthquake. Even though the civil-protection officials cited him
for instigating public alarm and panic and told him to stop, many
of the 75,000 residents had become increasingly uneasy.

Six members of the National Commission for Forecasting and Predicting Great Risks and the scientist heading the National Center on Earthquakes were ordered to a meeting on March 31, 2009, in L'Aquila by the head of the Department of Civil Protection. Their charge was to furnish citizens "with all the information available to the scientific community about the seismic activity of recent weeks." Those ordered to the meeting included a government official, the vice chairman of the Department of Civil Protection.

From this point on, there is confusion in the media about details. The scientists, who were used to working in closed sessions, were surprised to see a dozen government officials and other members of the public attending the meeting, which was apparently to be a one-hour review of the swarms of tremors. One of the scientists later was quoted in the minutes of the meeting as having said, "It is unlikely that an earthquake like the one in 1703 could occur in the short term, but the possibility cannot be totally excluded."

Only a week later, on the morning of April 6, 2009, disaster struck. A magnitude 6.3 earthquake jarred the town. While older people may have evacuated their homes in keeping with tradition, many others did not. In all, 309 people died, 1,500 were injured, and 65,000 were temporarily displaced. Twenty thousand buildings, many dating back to medieval times, were destroyed, partly as a result of liquefaction caused by the earthquake.

In June of 2012, government prosecutors charged the scientists and government official with negligence and manslaughter for failing to adequately evaluate and communicate the potential risk to the population. They were convicted and sentenced to six years in prison, fined about $400,000 each, and disqualified from ever holding public office. The appeal process for the defendants could last until 2019.

One resident said the following of the scientists: "Either they

FIGURE 10.1 Collapsed government building after
the L'Aquila earthquake in 2009. *By TheWiz83 from
it.wikipedia, and reproduced here under the Creative
Commons Attribution-Share Alike 3.0 Unported License.*

didn't know certain things, which is a problem, or they didn't
know how to communicate what they did know, which is also a
problem." The case has had a chilling effect on the willingness
of scientists to share their expertise with the public. Researchers around the world, normally wary of being misquoted by the
news media, are more wary than ever that journalists or public
officials will err when they translate their risk analyses for public consumption. If scientists fear reprisal in the event that they
communicate poorly, they may choose not to communicate at all,
which would be a tragedy in terms of minimizing the effects of
hazards on the public. Ideas are even emerging that attempting to
predict earthquakes may do more harm than good.[2]

This L'Aquila case illustrates many of the matters facing scientists and the public that have been covered in this book, particularly in the "Reflections" sections of previous chapters: What do we know and what don't we know about a particular disaster situation? In other words, what are the known knowns and known unknowns? How do scientists communicate to the public about events that have a low probability but very high consequences, be they earthquakes, landslides, volcanic eruptions, tsunamis, hurricanes, or tornadoes? How do we educate the public about the changing nature of scientific conclusions as new evidence is presented for examination? How can scientists interact most productively with the press? With politicians and other community leaders? At the time of this writing the answers are in flux.

RISK IN THE MODERN WORLD

Those of us who have lived through the second half of the twentieth century into the twenty-first century have witnessed a profound transition in the biological and physical relationship between humans and the rest of the planet. In the beginning of the twentieth century, our planet still had real islands: there were frontiers that held new lands, mysteries, adventures, cultures, and resources. Until recently, disasters—even megadisasters—seemed to be events that happened "somewhere else." By the end of the twentieth century, these islands and frontiers had become merged into a relatively seamless planet by science and technology, global communication, and an expanding population. We are crowding 7 billion people, and counting toward more than 9 billion people by 2050, onto this planet—many living in places in danger from natural disasters. Our sheer numbers ensure that natural disasters in the future will exacerbate the type and magnitude of natural disasters, because we will be crowding more people into areas of risk.

It no longer takes the 100-year flood or massive volcanic eruption to disrupt our economies and way of life. An eruption the size of the 2010 Eyja eruption occurs somewhere on the planet every few years, and the Eyja eruption showed how a relatively small one can have economic repercussions around the globe. As shown by the events at Eyja and then again at Tohoku in 2011, a natural event of a size that occurs roughly every few years or decades can disrupt global trade and productivity for several years, if not longer, and disrupt personal lives forever. Now an ever-increasing portion of the population lives on lands exposed to natural disasters. The challenge of maintaining civilized societies on the planet is becoming increasingly complex.

In this context, the role of scientists, especially earth scientists, has evolved. Although still a subject of discussion and debate, scientists who work with natural processes that may lead to disasters can no longer isolate themselves from the societal consequences of their work. It is incumbent on them to use their knowledge wisely on behalf of the public who may be affected by the processes that they study and to assist in communicating that knowledge.[3]

Civilization is a fragile enterprise: we depend on a favorable global climate, abundance of natural resources, and geological as well as social stability. But geology is not always stable, and a normally innocuous, even lovely, landscape can change states nearly instantaneously to pose a threat. We are vulnerable to disasters when instability sets in. The fragility of our societies is compounded by our propensity to take for granted our seemingly stable planet and its resources—and each other—with our current political systems and a shortsighted view of the future.

By nature, humans are much less concerned with issues on long timescales than on the timescales of ourselves, our children, and, perhaps, our grandchildren. This tendency toward a relatively short-term outlook makes planning for a 50-year timescale difficult—and for the 100-year timescale, nearly impossible—especially on the global scale that would be necessary to ameliorate the impact of these inevitable natural events.

To see this, we have only to look at the complex issues faced by the Japanese people, leaders, scientists, and engineers as they continue to recover from the 2011 earthquake and tsunami. What do they do in their crowded country with fertile, habitable land that will certainly, on these decadal and century timescales, experience another disaster? The Tohoku earthquake and tsunami led to the shutdown of all nuclear reactors on their islands, and a resolve to go nuclear-free. Barely more than a year later, as the hot, humid days of summer loomed ahead, Japan's prime minister, Yoshihiko Noda, said, "Japanese society cannot survive if we stop all nuclear reactors or keep them halted."[4] As the Japanese look back at the experience of the people of the Philippines with the 1991 eruption of Pinatubo, or the people of Haiti recovering from their 2010 earthquake, the most outstanding impression they will gain is that the task is daunting. The Japanese have been, and will continue to be, pioneering a new dynamic—that of coexistence with natural disasters on islands with finite resources. The regions of the world in which such a dynamic must be solved are increasing in size and number.

Natural and stealth disasters occur because energy in a system changes forms—sometimes catastrophically, as in many natural disasters, and sometimes gradually, as in many stealth disasters. The change in the form of energy causes a change of state in the system. Sometimes these changes occur without significant human influence, other times with major human influence. It behooves us to learn where and when our influence is significant and what direction it is taking a system. Even though the geological processes may be beyond our control, we can moderate our actions and choose behaviors that do not obviously lead to harm. Abiding by the precautionary principle and avoiding causing harm to the public or environment, even in the absence of scientific consensus, is wise and humane.

A call for action on natural and stealth disasters is not new; the United Nations General Assembly, for example, declared the 1990s as the International Decade for Natural Disaster Reduction. However, it is urgent that this call be repeated loudly, clearly, and widely because of the increasingly large number of people who live under threats of natural disasters, and the growing human and economic costs as we push the limits of the livable habitats on this planet. Those exposed to natural disasters include the most impoverished on our planet, and it is our moral obligation not only to protect ourselves and those close to us but to reach out far beyond our borders in thinking and in action. In the next section, I will show how one small group of us thought about the problem.

PROPOSAL FOR A CDC FOR PLANET EARTH

In many situations related to disasters, humans have two choices: do nothing, or remediate and adapt to the processes that Earth gives us. Assuming that we do not want to do nothing, and that we choose to remediate and adapt, a large body of wise leadership; huge databases for research, education, and policy; and the personal commitment of individuals around the planet will be required at a global scale.

In the Preface and Chapter 1, I posed a number of questions, and it's time to address at least a few of them here: How are natural disasters in the future going to compare with those of the past in impacting our lives on this ever-more-crowded planet? Is it possible that our sheer numbers will exacerbate the type and magnitude of natural disasters? What do we collectively know? What small fragment of that collective knowledge do I know? How can I extract knowledge from the collective wisdom to reduce the number of unknown unknowns in my life? How can I insert what I know into the collective framework so that it may become useful to others?

I have had the great privilege of working with four colleagues and friends in the geological community for nearly three decades about critical issues faced by our societies: Ward Chesworth, Pete Palmer, Paul Reitan, and E-an Zen.[5] With these four "wise men," my mentors, I spent many hours, days, and years pondering the relation of humans to planetary resources, to natural disasters, and to the new and insidious stealth disasters that are creeping up on us. During the time we were thinking about these problems, the epidemic of SARS in 2002–03 and the potential epidemic of bird flu in 2005 dramatically illustrated the global interconnectedness and fragility of our species. Without rapid detection and acknowledgment of the SARS problem, as well as aggressive monitoring and treatment through such organizations as the Centers for Disease Control and Prevention (CDC) and the World Health Organization (WHO), disastrous pandemics of SARS or flu might have occurred (and still may). Monitoring showed that, in the absence of initial control measures, individual SARS spreaders infected about three people on the average in Hong Kong and Singapore. As control measures were instituted, the transmission rate fell. With four key actions—detection, acknowledgment, aggressive monitoring, and treatment—the pandemic was halted, at least temporarily, in the societies that instituted these remedial measures. These events, and those actions, triggered in our discussions many of the ideas that I will describe next.

We thought about the analogy between the SARS and bird flu outbreaks and our natural disasters. A major similarity is that earth scientists know—as did epidemiologists about SARS and bird flu—a lot about the problems and the causes of the disasters. Earth scientists understand the dynamics of these disasters. There is also ample evidence for almost anyone to see how we humans are—wisely or stupidly—dealing with these dynamics. This knowledge has been compiled by individuals from many disciplines and subdisciplines who have put their individual building blocks of experience into the edifice of human knowledge in a variety of venues, including publication in scholarly journals, books, the World Wide Web, and service in private and public sectors.

Lack of knowledge is not our problem. Increasingly, this knowledge is easily available to any member of the public who desires to access it. A good starting place for individuals in the US is to do an Internet search for the geological survey in their state (for example, the Illinois State Geological Survey) or for the national US Geological Survey, which has district offices across the country dealing with geological issues at all scales. Governments of many other nations and their states or provinces have similar materials.

Let me give an example of what you might find if you live in an area prone to earthquakes, volcanic eruptions, or floods in the US. You would find documents, maps, and, most probably, advice. You would find geologists, and perhaps even a few engineers, to talk to about your situation. Generally, these agencies have had in place, for quite a long time, mechanisms for making their work available to the public. They also have procedures for ensuring that the material that they are officially putting out is as accurate as possible. In my decade of working for the US Geological Survey, every publication—whether as small as an abstract submitted to a formal scientific meeting, a map, a paper submitted to a peer-reviewed journal, or material prepared for the public—had to undergo internal peer review within the agency and have "Director's approval"[6] before being sent out to a scientific journal (where it would then undergo peer review from the external community) or to the public. Many of us viewed these internal procedures as draconian, but they generally ensured a high-quality, accurate, and credible product.

Gifted scientists sometimes are able to connect with and inspire the public, but not always are they the best ones for this task or for other tasks required to prevent or minimize the consequences of disasters. We scientists are generally not civil engineers who can design and implement solutions to vulnerabilities that we find. And neither scientists nor engineers are generally in a position to make policy, although there are examples of scientists and engineers who have successfully blended or made the transition to policy.

To me, the events at L'Aquila are tragic for both the residents and the scientists involved. Something went wrong along the way as the tragedy unfolded. The people and prosecutors of L'Aquila understand that earthquakes cannot be predicted at this time; that is not the basis of the manslaughter charge. Rather, there appears to have been a breakdown in communication, planning, implementing, and, particularly, formulating and enforcing construction practices that could stand up to the obvious earthquake dangers in the area. The events at L'Aquila confirm, retrospectively, a number of generalizations.

In dealing with future disasters globally, different sectors encompassing a substantial part of the global population will be required to deal with the problem: *scientists* to provide impartial facts and uncertainties; *engineers* to propose and implement technical solutions; *financiers* to manage the costs; *negotiators* to balance the realities of political, economic, religious, and cultural values; *facilitators, communicators, and enforcers* who are responsive to all of the inputs and can make sure that the recommendations, once agreed upon, are communicated, implemented, and enforced; and, perhaps above all, *teachers and learners.*

Here are a few examples of existing bodies that might, if coordinated, provide a template for an integrated way to deal with natural disasters:

Scientists and engineers. In the US, representatives of the National Academy of Sciences, the National Academy of Engineering, and the Institute of Medicine now routinely meet with their counterparts around the world to discuss issues of concern under their charters. These are not restricted to, but include, natural disasters.

Financiers. In the nongovernmental sector of many countries, the costs of disasters are currently dealt with primarily through insurance. Unfortunately, all too many people in the world have no access to insurance or simply cannot afford it. Although insurance is an ancient concept, modern insurance—a principal mechanism for coping with disaster through insurance and reinsurance—began after the Great Fire of London in 1666. Yet the sobering events of the early twenty-first century showed vulnerabilities in the existing structure of the insurance industry that must be remedied.

Negotiators. Two examples of global attempts to deal with disasters through negotiation demonstrate both promise and problems. The 1989 international Montreal Protocol on ozone was an outstandingly successful accord for reducing dangerous ozone levels in the atmosphere. If the agreement is adhered to, the ozone layer will be restored to natural levels by 2050. As fraught with controversy and publicity as it is, the workings of Intergovernmental Panel on Climate Change (IPCC) represent a sustained attempt by leaders in all fields to address the important contemporary issue of climate change with copious public input from all sectors.

Facilitators, communicators, and enforcers. One example of scientists communicating successfully to the public comes from the meteorological community, with its array of weather forecasters speaking directly to the public. Generally, however, the communication falls on nonscientists, and that can give rise to problems like those at L'Aquila. Clear and appropriate communication of risk to officials and to the public is essential, because only then can we expect rational action to follow. For example, citizens, zoning commissions, and construction workers must understand the reason for appropriate building codes and their enforcement. Sometimes communication can be accomplished through government agency spokespersons, other times through civil defense institutions. Routine enforcement, such as of building codes, requires inspectors who act with integrity and diligence. The US National Guard and the Canadian Peacekeepers, with their roles as political and human mediators in complex state, national, and global problems, may provide models for the enforcement function.

Teachers and learners. Here I would include every level of teaching and learning, from formal to informal, to personal exchanges, community meetings, and rapidly developing new outlets such as Wikimedia and Coursera. We are all teachers, and we can all be learners. The model that Californians have for earthquake disaster preparedness is one that the world would do well to examine. The Internet ensures that even the farthest reaches of the globe can be reached with information. Without personnel enlightenment, wise choices and actions are nearly impossible.

To organize our thinking further and even speculate about an organization that could address our perceived global needs, we ("the wise men + 1," as I have dubbed us) examined the mandates of a number of organizations, and ended up taking the mandates of the US Centers for Disease Control and Prevention (CDC) as a model for our thinking. We created a hypothetical global body that might address hazards, and nicknamed it the "CDC-PE"—the Center for Disaster Control for Planet Earth.

To paraphrase the CDC mandate, and to think about it on a global scale instead of a national scale, we propose that:

The (hypothetical) CDC-PE should be recognized as the lead world body for protecting the long-term health and safety of the planet and all of its inhabitants, providing credible information to enhance decisions relating to all resources and processes of the planet, and promoting wisdom in living in harmony with all elements of the planet through strong international cooperation. The CDC-PE serves as the international focus for developing and applying conservation concepts, and promoting educa-

tion activities designed to improve the conditions for continued human existence on the planet.

Again, loosely following the CDC, we propose that in carrying out its activities, the CDC-PE might have the following six core functions:

1. *To manage information by assessing data about our planetary processes, resources, and hazards, including the uncertainties in the data and our knowledge; assessing trends; and stimulating and setting the agenda for research and development.*

2. *To set, validate, monitor, and pursue the proper implementation of standards—for example, the use of resources, the disposal of waste, the use of land and human modifications of land—especially in areas exposed to geological hazards.*

3. *To catalyze change through technical and policy support that stimulates cooperation and action and helps to build sustainable global capacity.*

4. *To negotiate and sustain local, national, and global partnerships.*

5. *To articulate consistent, ethical, and evidence-based policy and advocacy positions.*

6. *To provide the best scientific data about the planet, its resources and its hazards, in a form that individuals, policy makers, educators, and learners can use.*

The magnitude of the task, the global scope, and the long time-scales involved require the resources of long-lived institutions to deal with disasters, for in them we can vest our knowledge with hopes for continuity at the large scale, and for the long duration, required. And here lies the answer to the last of the questions I asked earlier in the book and repeated in this chapter: With modern technology, it is possible to insert what we individually know into the collective framework of knowledge so that it may become useful to others, and we can act on this knowledge to create better and safer living on the planet. Every action matters in some way—sometimes to the negative, sometimes to the positive, sometimes with little consequence, and sometimes with enormous consequence.

The CDC-PE was simply an abstract concept to guide our thinking, and we concluded that it must be (1) global, (2) credible, (3) scientifically based, and (4) sensitive to political, economic, religious, and cultural values. The United Nations has, in fact, an Office for Disaster Risk Reduction that aims to accomplish these goals, but even they, through their Scientific and Technical Committee, have stated that better mechanisms are needed for integrating science and technology into policy and practice. Much remains to be done. Individually, we can start by applying these concepts to our lives at whatever local levels we can ("think global, act local," "do unto your neighbor as you would have done unto you") and abiding by the ancient wisdom that "an ounce of prevention is worth a pound of cure"—that is, live by the precautionary principle. If we have the collective will to think about ourselves and our relation to the planet at the scale of a "CDC for Planet Earth," we can greatly reduce the impact of disasters on us, as well as our impact on the resources of the planet.

ACKNOWLEDGMENTS

I thank my editor Jack Repcheck, Devon Zahn, Hannah Bachman, Don Rifkin, and the Norton staff, and Stephanie Hiebert, freelance copyeditor, for their diligence and patience through the creation of this book. Jack may have thought that he knew how to handle first-time authors (by requiring an extensive outline that took me almost a year to get by him). He may not, however, have accounted for my inability to stick with the outline! For his patience and guidance, I am very grateful. I thought that I understood *The Chicago Manual of Style* until I saw the editing of my manuscript by Stephanie and others at Norton. For their help and improvements to the manuscript, I am truly grateful.

I have discovered and rediscovered several times in my career the generosity of fellow academic geoscientists in sharing data, ideas, and concerns. A most rewarding discovery in the process of writing this book was that this sharing and helpfulness exists throughout the scientific and nonscientific communities that I reached out to—not only geoscientists whom I had not previously known, but also physicists, chemists, people in the publishing industry, amateur scientists, nature lovers, geocachers, and other people around the globe whom I did not know but approached by email. Many went out of their way to provide high-resolution images, find images in their photo collections or archives, draft special versions of their work, or suggest alternatives. To those whose figures I used, and those whose figures I could not use because of either space or image resolution considerations, I convey my gratitude.

Finally, most important to me have been those to whom this book is dedicated. Without the wisdom of those I call my "wise men," who nurtured my career through four decades with support, constructive criticism, and grounding of my sometimes wild ideas, and the love, support, and humorous perspective on this book endeavor from my husband, Gerard Lopez, none of it would have happened.

NOTES

Preface

1. *Merriam-Webster's Collegiate Dictionary*, "disaster" entry, accessed February 10, 2013, http://www2.merriam-webster.com/cgi-bin/mwdictsn?book =Dictionary&va=disaster.
2. The word "tsunami" is, technically, both the singular and plural form. However, following common English practice, I use the word "tsunamis" for the plural form.

Chapter 1 GEOLOGIC CONSENT—DO WE HAVE IT OR NOT?

1. W. Durant, "What Is Civilization?," *Ladies Home Journal*, January 1946.
2. T. Frank, "'Disasters' Strain FEMA's Resources," *USA Today*, October 24, 2011, http://www.usatoday.com/news/washington/story/2011-10-23/disasters -strain-fema-funds/50886370/1.
3. Ibid.
4. J. L. Warren and S. Kieffer, "Risk Management and the Wisdom of Aldo Leopold," *Risk Analysis* 30, no. 2 (2010): 165–74.
5. See, for example, S. W. Kieffer, P. Barton, W. Chesworth, A. R. Palmer, P. Reitan, and E. Zen, "Megascale Processes: Natural Disasters and Human Behavior," in *Preservation of Random Megascale Events on Mars and Earth: Influence on Geologic History*, Geological Society of America Special Papers 453, ed. M. G. Chapman and L. P. Keszthelyi (Boulder, CO: Geological Society of America, 2009), 77–86.
6. World Bank and United Nations, *Natural Hazards, UnNatural Disasters:*

The Economics of Effective Prevention (Washington, DC: World Bank, 2010).

7. In its publication the World Bank uses the terms "hazard" and "disaster" in a different way than how I use them in this book.

8. US Department of Defense, "DoD News Briefing—Secretary Rumsfeld and Gen. Myers," February 12, 2002, http://www.defense.gov/Transcripts/Transcript.aspx?TranscriptID=2636.

9. "Foot in Mouth Award: Past Winners," Plain English Campaign, accessed December 10, 2012, http://www.plainenglish.co.uk/awards/foot-in-mouth-award/foot-in-mouth-winners.html.

10. From the speech Socrates gave while defending himself on trial, as documented by Plato in the *Apology.*

11. I ignore here the category of "unknown knowns." For more on this topic, see, for example, discussions by the philosopher Slavoj Zizek as expounded in the filmed lecture *Slavoj Zizek: The Reality of the Virtual*, a 2004 film by Ben Wright.

12. D. K. Chester, "The 1755 Lisbon Earthquake," *Progress in Physical Geography* 25, no. 3 (2001): 363–83.

13. N. Taleb, *The Black Swan: The Impact of the Highly Improbable*, 2nd ed. (New York: Penguin, 2010). In this context, a black swan is an event that has a low probability but very high consequences. Human society is often unprepared for such events—for example, the attacks of September 11, 2001, or the Tohoku earthquake and tsunami—but in hindsight they appear to have been predictable incidents within the appropriate context. The objective of coping with black swans, or turning them "white," as a society is to identify and address areas of vulnerability.

14. The ongoing lack of disaster preparedness has been discussed in a number of recent books, including Amanda Ripley's *The Unthinkable: Who Survives When Disaster Strikes and Why* (New York: Crown, 2008), and Nassim Taleb's discussion of the disproportionate impact of rare events on both people and their institutions in *The Black Swan.*

15. I emphasize events that have happened recently rather than far back in the geological record—typically within the last tens of thousands of years, compared to geological time that stretches back more than 4 billion years.

16. S. W. Kieffer, "Geology: The Bifocal Science," in *The Earth Around Us: Maintaining a Livable Planet,* ed. J. Schneiderman (San Francisco, Freeman Press, 2000), 2–17.

Chapter 2 DYNAMICS AND DISASTERS

1. J. Samenow, "NOAA: 2011 Sets Record for Billion Dollar Weather Disasters in the U.S.," *Washington Post*, December 7, 2011, http://www.washingtonpost .com/blogs/capital-weather-gang/post/noaa-2011-sets-record-for-billion-dollar-weather-disasters/2011/12/07/gIQAjD9kcO_blog.html.

2. Aftershocks and earthquakes on nearby faults, including one of magnitude 6.3 that caused 179 casualties as buildings that had been preweakened by the September 2010 earthquake collapsed, continued to rock the weakened infrastructure of the city for months.

3. Pronounced "AYE-ya-FYAH-dla-JOW-kudl," with the capitalized syllables more heavily accented. F. Sigmundsson, S. Hreinsdóttir, A. Hooper, T. Árnadóttir, R. Pedersen, M. J. Roberts, N. Óskarsson, et al., "Intrusion Triggering of the 2010 Eyjafjallajokull Explosive Eruption," *Nature* 468 (2010): 426–30.

4. Oxford Economics, "The Economic Impacts of Air Travel Restrictions Due to Volcanic Ash: A Report Prepared for Airbus, 2010," accessed December 3, 2010, http://www.oxfordeconomics.com/OE_Cons_Aviation.asp.

5. BBC News, "Ash Chaos: Row Grows over Airspace Shutdown Costs," April 22, 2010, http://news.bbc.co.uk/2/hi/uk_news/8636461.stm. The following earlier paper considered the changing nature of regional-scale effects: D. M. Johnston, B. F. Houghton, V. E. Neall, K. R. Ronan, and D. Paton, "Impacts of the 1945 and 1995–1996 Ruapehu Eruptions, New Zealand: An Example of Increasing Societal Vulnerability," *Geological Society of America Bulletin* 112 (2000): 720–26.

6. Technically, in an *isolated* system.

Chapter 3 WHEN TERRA ISN'T FIRMA

1. Although the Richter magnitude scale introduced in 1935 is well known, it was replaced in the 1970s by the "moment magnitude" scale. The Richter scale is a logarithmic base-10 scale obtained from the amplitude of waves measured on a seismograph. The moment magnitude scale, introduced to address deficiencies in the Richter scale for large earthquakes, is obtained from estimates of the rigidity of the rocks, the area of the fault rupture, and

the displacement distance. In either scale, an increase of 1 on a logarithmic scale corresponds to an increase by a factor of 10 in the amplitude of the waves, and an increase of 31.6 (the square root of 1,000) in the energy released. In this book, earthquake magnitudes are moment magnitudes where possible.

2. Survivor Frantz Florestal, as reported in J. Sturcke, "Haiti Earthquake: Survivors' Stories," *Guardian*, January 14, 2010, http://www.guardian .co.uk/world/2010/jan/14/haiti-earthquake-survivors.

3. The official death toll reported by the Haitian government is 316,000. From surveys of households, Kolbe et al. reported 158,679 dead (A. R. Kolbe, R. A. Hutson, H. Shannon, E. Trzcinski, B. Miles, N. Levitz, M. Puccio, L. James, J. R. Noel, and R. Muggah, "Mortality, Crime and Access to Basic Needs before and after the Haiti Earthquake: A Random Survey of Port-au-Prince Households," *Medicine, Conflict, and Survival* 26, no. 4 [2010]: 281–97), and a USAID report suggested that the death toll might be as low as 46,000. For an overview of the difficulties of estimating deaths accurately, see M. R. O'Conner, "Two Years Later, Haitian Earthquake Death Toll in Dispute," *Columbia Journalism Review: Behind the News*, January 12, 2012, http://www.cjr.org/behind_the_news/one_year_later_haitian_earthqu .php?page=all.

4. The Christchurch earthquake is sometimes referred to as the "Canterbury earthquake." The region of Canterbury is the largest region on New Zealand's South Island; Christchurch is Canterbury's largest city.

5. The statistics here are taken from the US Geological Survey, "Historic World Earthquakes: China," accessed December 23, 2012, http://earthquake .usgs.gov/earthquakes/world/historical_country.php#china. The true extent of disasters in China is fairly uncertain because, until recently, the political environment in China made it difficult to assess the statistics reported by various regional units. In 1920, the magnitude 7.8 Haiyuan earthquake (also called the 1920 Gansu earthquake) killed over 200,000 people, possibly as many as 273,000. In 1976, the Great Tangshan Earthquake of magnitude 7.5 killed hundreds of thousands (estimates range from 242,000 to 700,000). In 2008, over 87,000 people died in the magnitude 7.9 Szechuan (or Sichuan) earthquake, and nearly 3,000 were killed in the magnitude 6.9 earthquake in Yushu in 2010.

6. It is also equivalent to all the people estimated to have been alive on the planet shortly after the ice ages ended.

7. For a more detailed discussion of earthquakes, see, for example, R. S. Yeats, K. Sieh, and C. R. Allen, *Geology of Earthquakes* (New York: Oxford University Press, 1997).

8. R. H. Sibson, "Brecciation Processes in Fault Zones: Inferences from Earthquake Rupturing," *Pure and Applied Geophysics* 124, nos. 1–2 (1986): 159–75.

9. "Bows and Strings," University of New South Wales, accessed December 23, 2012, http://www.phys.unsw.edu.au/jw/Bows.html. The reason that a violinist puts rosin on the horsehair bow is to increase the friction between the bow and string so that the bow drags the string for a while before it slips and vibrates to produce the tone.

10. H. Kanamori and E. Brodsky, "The Physics of Earthquakes," *Physics Today* 54, no. 6 (2001): 34–40.

11. D. M. Russell, "How Long Was the Haiti Earthquake from Jan 12, 2010?," *SearchReSearch* (blog), February 11, 2010, http://searchresearch1 .blogspot.com/2010/02/answer-how-long-was-haiti-earthquake.html.

12. R. Bilham, "Lessons from the Haiti Earthquake," *Nature* 463 (2010): 878–79.

13. T. Lay and H. Kanamori, "Insights from the Great 2011 Japan Earthquake," *Physics Today* 64, no. 12 (2011): 33–39.

14. An excellent source of information about this quake for general science readers is at http://outreach.eri.u-tokyo.ac.jp/eqvolc/201103_tohoku/eng. The specifics reported here come from C. J. Ammon, T. Lay, H. Kanamori, and M. Cleveland, "A Rupture Model of the 2011 off the Pacific Coast of Tohoku Earthquake," *Earth, Planets, and Space* 63 (2011): 693–96.

15. An excellent Web page with animations of this material provided by Miaki Ishii is at http://www.seismology.harvard.edu/research_japan.html.

16. 750–800 miles long and 125 miles wide.

17. S. Stein and E. A. Okal, "Speed and Size of the Sumatra Earthquake," *Nature* 434 (2005): 581–82; M. Ishii, P. M. Shearer, H. Houston, and J. E. Vidale, "Extent, Duration and Speed of the 2004 Sumatra-Andaman Earthquake Imaged by the Hi-Net Array," *Nature* 435 (2005): 933–36.

18. Famously, Mao Tse-tung and the Communists lived in these dwellings from 1935 to 1948 while leading resistance against the Japanese in the Second Sino-Japanese War and consolidating his hold over the Communist Party.

19. Amazing footage of this process taking place in a park in Tokyo during the 2011 Tohoku earthquake can be found at http://news.blogs.cnn

.com/2011/03/14/gotta-watch-ireporters-capture-scope-of-quake/?hpt=C2 (accessed September 2, 2012). Dry ground suddenly becomes wet, puddles appear where none had been minutes before, and small geysers of water erupt a foot into the air through cracks that open and close as the seismic waves propagate through the park.

20. This description is a composite from eyewitness accounts of the 1811–12 New Madrid quakes and the 1906 San Francisco quake. Additional information comes from L. Bringier, "Notices of the Geology, Mineralogy, Topography, Production, and Aboriginal Inhabitants of the Regions around the Mississippi and Its Confluent Waters," *American Journal of Science* 3 (1821): 15–46. The US Geological Survey has estimated that these were between magnitude 7.8 and 8.4, which would make them bigger than any earthquakes in the history of the mainland US (Alaska has had a bigger quake). However, Seth Stein has argued that these estimates are not correct and, along with others, he claims that they were between magnitude 6.8 and 7.0. A comprehensive description of the New Madrid events and of modern controversy about the probability of future quakes there can be found in S. Stein, *Disaster Deferred: How New Science Is Changing Our View of Earthquake Hazards in the Midwest* (New York: Columbia University Press, 2010).

21. There may be oral traditions within the Native American communities of this region, such as the Chickasaw. See, for example, "The Legend of Reelfoot Lake," accessed December 22, 2012, http://www.ecsis.net/dsv/lakecounty/reelfoot/legend.html.

22. C. A. von Hake, "Missouri: Earthquake History," *Earthquake Information Bulletin* 6, no. 3 (May–June 1974), http://earthquake.usgs.gov/earthquakes/states/missouri/history.php.

23. D. Zhang and G. Wang, "Study of the 1920 Haiyuan Earthquake-Induced Landslides in Loess (China)," *Engineering Geology* 94 (2007): 76–88.

24. I. Thono and Y. Shamoto, "Liquefaction Damage to the Ground during the 1983 Nihonkai-Chubu (Japan Sea) Earthquake in Aomori Prefecture, Tohoku, Japan," *Natural Disaster Science* 8, no. 1 (1986): 85–116.

25. N. N. Ambraseys, "Engineering Seismology," *Earthquake Engineering and Structural Dynamics* 17 (1988): 1–105.

26. B. H. Fatherree, "New Vistas in Civil Engineering, 1963–1980: Soil Mechanics and Earthquake Engineering," in *The History of Geotechnical Engineering at the Waterways Experiment Station 1932–2000* (US Army

Corps of Engineers, 2006), chap. 10, http://gsl.erdc.usace.army.mil/gl-hi story/Chap10.htm.

27. J. Broughton and R. Van Arsdale, "Surficial Geologic Map of the Northwest Memphis Quadrangle, Shelby County, Tennessee, and Crittenden County, Arkansas," *U.S. Geological Survey Scientific Investigations Map* 2838 (2004).

28. Reported in D. Finkelstein and J. R. Powell, "Lightning Production in Earthquakes" (paper presented at the 15th General Assembly, International Union Geodesy and Geophysics, Moscow, 1971).

29. J. S. Derr, "Earthquake Lights: A Review of Observations and Present Theories," *Bulletin of the Seismological Society of America* 63 (1973): 2177–87.

30. A YouTube video at http://www.youtube.com/watch?v=f14pQakxXjc (accessed April 23, 2013) records these fantastic lights during the 2007 Pisco, Peru, magnitude 8.0 earthquake.

31. In fault zones, pseudotachylites are sometimes referred to as "mylonites."

32. A. Lin, *Fossil Earthquakes: The Formation and Preservation of Pseudotachylytes*, Lecture Notes in Earth Sciences 111 (Berlin: Springer, 2007), 38.

33. S. J. Shand, "The Pseudotachylyte of Parijs (Orange Free State), and Its Relation to 'Trap-Shotten Gneiss' and 'Flinty Crush-Rock,'" *Geological Society of London Quarterly Journal* 72 (1916): 198–221.

34. Ibid.

35. See review by J. Spray, "Pseudotachylyte Controversy: Fact or Fiction?," *Geology* 23 (1995): 1119–22.

36. Ibid.

37. Even minerals that are not black themselves can appear black when they are submicroscopic nanograins.

38. Technically, the duration of these magnetic fields is the length of time it takes the material to cool to the temperature at which the magnetic field is no longer recorded—a temperature called the Curie point. For pure magnetite, this temperature is about 1,050°F–1,200°F, with the higher values occurring deeper in the earth at higher pressures.

39. F. Freund, M. A. Salgueiro da Silva, B. W. S. Lau, A. Takeuchi, and H. H. Jones, "Electric Currents along Earthquake Faults and the Magnetization of Pseudotachylite Veins," *Tectonophysics* 431 (2007): 131–41; E. C. Ferré, M. S. Zechmeister, J. W. Geissman, N. MathanaSekaran, and K. Kocak, "The Origin of High Magnetic Remanence in Fault Pseudotachylites: Theo-

retical Considerations and Implication for Coseismic Electrical Currents," *Tectonophysics* 402 (2005): 125–39.

40. D. McKenzie and J. N. Brune, "Melting on Fault Planes during Large Earthquakes," *Geophysical Journal of the Royal Astronomical Society* 29 (1972): 65–78.

41. Kanamori and Brodsky, "Physics of Earthquakes."

42. The most recent USGS maps are available as PDF files from the USGS: M. D. Petersen, A. D. Frankel, S. C. Harmsen, C. S. Mueller, K. M. Haller, R. L. Wheeler, R. L. Wesson, et al., "Seismic-Hazard Maps for the Conterminous United States, 2008," *US Geological Survey*, December 27, 2011, http://pubs .usgs.gov/sim/3195.

Chapter 4 THE FLYING CARPET OF ELM

1. An excellent description with illustrative figures of the sliding process can be found in O. Gregersen, *The Quick Clay Landslide in Rissa, Norway: The Sliding Process and Discussion of Failure Modes*, Norwegian Geotechnical Institute Publication 135 (Oslo: Norwegian Geotechnical Institute, 1981), http://www.ngi.no/upload/6485/Rissa_Quick_Clay_slide_NGI%20 Publ.135.pdf. The direct quotes in the text here are from this publication. The same material is graphically presented in the video, in English, by the Norwegian Geotechnical Institute at http://www.youtube.com/watch?v=3q-qfNlEP4A. The video includes not only footage of the events at Rissa, but also lab illustrations of the properties of quick clay. Especially enlightening is the illustration of turning the liquid quick clay back to a solid form by adding salt.

2. The slide started when a portion of the stockpiled dirt slid into the lake. Over a period of forty minutes, several minor slides followed toward the lake. Each minor slide resulted in a complete liquefaction of the quick clay, and the debris poured into the lake "like streaming water" (Gregersen, *Quick Clay Landslide in Rissa, Norway*). The affected area was initially only about 1,500 feet in length, and if the slide had stopped there it would have been just one of the many minor quick-clay slides that occur frequently in Norway. But at this point the bigger catastrophe started. A large flake of land, comprising about 9 acres (500 by 700 feet) started sliding into the lake. The major slides caused two or three significant tsunamis on Lake Botnen. These waves

propagated 3 miles across the lake, destroyed a sawmill and lumberyard, and caused significant damage in Leira, a small town on the far shore.

3. Gregersen, *Quick Clay Landslide in Rissa, Norway.*

4. See http://www.youtube.com/watch?v=3q-qfNlEP4A.

5. B. Lendon, "Family Dead in Basement after Sinkhole Swallowed Home," *CNN.com*, May 12, 2010, http://news.blogs.cnn.com/2010/05/12/family-found-dead-in-basement-after-sinkhole-ate-home.

6. A. Heim, "Der Bergsturz von Elm," *Deutsche Geologische Gesellschaft Zeitschrift* 34 (1882): 74–115, as reported in K. Hsü, "Catastrophic Debris Streams (Sturzstroms) Generated by Rockfalls," *Geological Society of America Bulletin* 86 (1975): 129–40.

7. W. G. Pariseau, "A Simple Mechanical Model for Rockslides and Avalanches," *Engineering Geology* 16, nos. 1–2 (1980): 111–23.

8. For a recent review of landslide models, see F. V. DeBlasio, *Introduction to the Physics of Landslides* (New York: Springer, 2011).

9. An excellent reference for landslide information such as these statistics is Dave Petley's *Landslide Blog* at http://blogs.agu.org/landslideblog. This particular information is from the September 13, 2012, post, accessed July 10, 2012.

10. F. C. Dai, C. F. Lee, J. H. Deng, and L. G. Tham, "The 1786 Earthquake-Triggered Landslide Dam and Subsequent Dam-Break Flood on the Dadu River, Southwestern China," *Geomorphology* 65 (2005): 205–21.

11. US Geological Survey, "3. Q: How Much Do Landslides Cost in Terms of Monetary Losses?," *Landslide Hazards Program: Frequently Asked Questions*, accessed July 10, 2012, http://landslides.usgs.gov/learning/faq/#q03.

12. L. Sahagun, "Walking Away from a Highway," *Los Angeles Times*, January 29, 2012, http://www.latimes.com/news/local/la-me-caltrans-high way39-20120129,0,2515708.story.

13. US Geological Survey, "3. Q: How Much?"

14. Some of this account is taken from J. J. Hemphill, "Assessing Landslide Hazard over a 130-Year Period for La Conchita, California" (paper presented at the Association of Pacific Coast Geographers Annual Meeting, September 12 and 15, 2001), http://www.geog.ucsb.edu/~jeff/projects/la_conchita/apcg2001_article/apcg2001_article.html.

15. R. W. Jibson, *Landslide Hazards at La Conchita, California*, Open-File Report 2005-1067 (US Geological Survey, 2005), http://pubs.usgs.gov/of/2005/1067/pdf/OF2005-1067.pdf. Also available at http://pubs.usgs.gov/of/2005/1067/508of05-1067.html#conchita05.

16. R. H. Campbell, *Soil Slips, Debris Flows, and Rainstorms in the Santa Monica Mountains and Vicinity, Southern California*, US Geological Survey Professional Paper 851 (Washington, DC: US Government Printing Office, 1975).

17. See http://www.youtube.com/watch?v=3q-qfNlEP4A.

18. D. N. Petley, N. J. Rosser, D. Karim, S. Wali, N. Ali, N. Nasab, and K. Shaban, "Non-seismic Landslide Hazards along the Himalayan Arc," in *Geologically Active: Proceedings of the 11th IAEG Congress. Auckland, New Zealand, 5–10 September 2010*, ed. A. L. Williams, G. M. Pinches, C. Y. Chin, T. J. McMorran, and C. I. Massey (London: CRC Press, 2010), 143–54.

19. S. Mir, "Clash between Police, Attabad Victims: Fresh Protests Erupt across G-B," *Express Tribune*, August 13, 2011, http://tribune.com.pk/story/230262/clash-between-police-attabad-victims-fresh-protests-erupt-across-g-b.

20. "China Land Minister Calls for Pre-Emptive Evacuation in Wake of Mudslides," *English.news.cn*, August 23, 2010, http://news.xinhuanet.com/english2010/china/2010-08/23/c_13456751.htm. Shipping through the massive Three Gorges Dam was halted as yet one more flood peak approached, bringing more than 45,000 cubic meters per second of water into the huge reservoir behind the dam ("Shipping through Three Gorges Dam Halted as New Flood Peak Approaches," *English.news.cn*, August 23, 2010, http://news.xinhuanet.com/english2010/china/2010-08/23/c_13458409.htm). Authorities anticipated that the flood peak could reach 56,000 cubic meters per second on August 24, 2010. This was the third suspension of shipping services through the dam during the summer of 2010.

21. D. N. Petley, "On the Initiation of Large Rockslides: Perspectives from a New Analysis of the Vajont Movement Record," in *Landslides from Massive Rock Slope Failure. Proceedings of the NATO Advanced Research Workshop on Massive Rock Slope Failure: New Models for Hazard Assessment. Celano, Italy, 16–21 June 2002*, NATO Science Series IV: vol. 49, ed. S. G. Evans, G. Scarascia Mugnozza, A. Strom, and R. L. Hermanns (Rotterdam, Netherlands: Kluwer, 2006), 77–84. An excellent summary can be found at "The Vaiont (Vajont) Landslide of 1963," *AGU Blogosphere*, December 11, 2008, http://blogs.agu.org/landslideblog/2008/12/11/the-vaiont-vajont-landslide-of-1963.

22. B. Voight and C. Faust, "Frictional Heat and Strength Loss in Some Rapid Landslides," *Geotechnique* 32 (1982): 43–54; B. Voight and C. Faust,

"Frictional Heat and Strength Loss in Some Rapid Landslides: Error Correction and Affirmation of Mechanism for the Vajont Landslide," *Geotechnique* 42 (1992): 641–43; C. R. J. Kilburn and D. N. Petley, "Forecasting the Giant, Catastrophic Slope Collapse: Lessons from Vajont, Northern Italy," *Geomorphology* 54 (2003): 21–32.

23. Hsü, "Catastrophic Debris Streams (Sturzstroms) Generated by Rockfalls," *Geological Society of America Bulletin* 86 (1975): 129–40. Excellent review in F. Legros, "The Mobility of Long-Runout Landslides," *Engineering Geology* 63 (2002): 301–31.

24. R. Shreve, *The Blackhawk Landslide*, Geological Society of America Special Papers 108 (Boulder, CO: Geological Society of America, 1968); R. Shreve, "Leakage and Fluidization in Air-Layer Lubricated Avalanches," *Geological Society of America Bulletin* 79 (1968): 653–58.

25. This velocity is reported in R. Shreve, "Sherman Landslide," *Science* 154 (1966): 1639–43.

26. Ibid.

27. Shreve, "Leakage and Fluidization"; Shreve, "Sherman Landslide."

28. Hsü, "Catastrophic Debris Streams"; G. S. Collins and H. J. Melosh, "Acoustic Fluidization and the Extraordinary Mobility of Sturzstorms," *Journal of Geophysical Research* 108 (2003): 2473–76; R. Han, T. Hirose, T. Shimamoto, Y. Lee, and J. Ando, "Granular Nanoparticles Lubricate Faults during Seismic Slip," *Geology* 39 (2011): 599–602; T. R. H. Davis, "Spreading of Rock Avalanche-Debris by Mechanical Fluidization, *Rock Mechanics and Rock Engineering* 15 (1982): 9–24.

29. T. Shinbrot, "Delayed Transitions between Fluid-Like and Solid-Like Granular States," *European Physical Journal* 22 (2007): 209–17; T. Shinbrot, N.-H. Duong, L. Kwan, and M. M. Alvarez, "Dry Granular Flows Can Generate Surface Features Resembling Those Seen in Martian Gullies," *Proceedings of the National Academy of Sciences of the USA* 101 (2004): 8542–46.

30. Hsü, "Catastrophic Debris Streams"; C. S. Campbell, "Self-Lubrication for Long Runout Landslides," *Journal of Geology* 97 (1989): 653–65.

31. H. J. Melosh, "The Physics of Very Large Landslides," *Acta Mechanica* 64 (1986): 89–99.

32. B. K. Lucchitta, "Valles Marineris, Mars—Wet Debris Flows and Ground Ice," *Icarus* 72 (1987): 411–29; Legros, "Mobility of Long-Runout Landslides"; K. P. Harrison and R. E. Grimm, "Rheological Constraints on Martian Landslides," *Icarus* 163 (2003): 347–62; F. V. De Blasio and A. A.

Elverhøi, "A Model for Frictional Melt Production beneath Large Rock Avalanches," *Journal of Geophysical Research* 113 (2008): F02014; K. N. Singer, W. B. McKinnon, P. M. Schenk, and J. M. Moore, "Massive Ice Avalanches on Iapetus Mobilized by Friction Reduction during Flash Heating," *Nature Geoscience* 5 (2012): 574–78.

33. B. Voight and J. Sousa, "Lessons from Ontake-san: A Comparative Analysis of Debris Avalanche Dynamics," *Engineering Geology* 38 (1994): 261–97.

34. E. C. Beutner and G. P. Gerbi, "Catastrophic Emplacement of the Heart Mountain Block Slide, Wyoming and Montana, USA," *Geological Society of America Bulletin* 117 (2005): 724–35.

35. J. P. Craddock, D. H. Malone, J. Magloughlin, A. L. Cook, M. E. Rieser, and J. R. Doyle, "Dynamics of the Emplacement of the Heart Mountain Allochthon at White Mountain: Constraints from Calcite Twinning Strains, Anisotropy of Magnetic Susceptibility, and Thermodynamic Calculations," *Geological Society of America Bulletin* 121 (2009): 919–38.

36. This discussion is excerpted from S. W. Kieffer, "Geology: The Bifocal Science," in *The Earth Around Us: Maintaining a Livable Planet*, ed. J. Schneiderman (San Francisco, Freeman Press, 2000), 2–17.

Chapter 5 THE DAY THE MOUNTAIN BLEW

1. As reported by Harry Kaiakokonok, eyewitness of the Katmai/Novarupta eruption of 1912, in "Witness: Firsthand Accounts of the Largest Volcanic Eruption in the Twentieth Century," *National Park Service*, April 2004, http://www.nps.gov/katm/historyculture/upload/Witnessweb.pdf.

2. Pliny the Younger, "Letters 6.16 and 6.20," from Penguin translation by Betty Radice, accessed December 29, 2012, http://www.u.arizona.edu/~afutrell/404b/web%20rdgs/pliny%20on%20vesuvius.htm.

3. G. Mastrolorenzo, P. Petrone, L. Pappalardo, and M. F. Sheridan, "The Avellino 3780-Yr-B.P. Catastrophe as a Worst-Case Scenario for a Future Eruption at Vesuvius," *Proceedings of the National Academy of Sciences of the USA* 103 (2006): 4366–70.

4. K. Barnes, "Volcanology: Europe's Ticking Time Bomb," *Nature* 473 (2011): 140–41.

5. The name is actually not the name of the volcano, but the name of the small glacier that caps the underlying volcanic mountain.

6. S. Self, J.-X. Zhao, R. E. Holasek, R. C. Torres, and A. J. King, "The Atmospheric Impact of the 1991 Mount Pinatubo Eruption," in *Fire and Mud: Eruptions and Lahars of Mount Pinatubo, Philippines*, ed. C. G. Newhall and R. S. Punongbayan (Quezon City: Philippine Institute of Volcanology and Seismology, 1996).

7. S. C. Singh, G. M. Kent, J. S. Collier, A. J. Harding, and J. A. Orcutt, "Melt to Mush Variations in Crustal Magma Properties along the Ridge Crest at the Southern East Pacific Rise," *Nature* 394 (1998): 874–78.

8. Technically, called "phreatic" eruptions after the Greek word for "well" or "spring." In modern use, the word applies to groundwater in general.

9. A. Lacroix, *La Montagne Pelée et ses eruptions* (Paris: Masson et Cie, 1904).

10. S. W. Kieffer, "Blast Dynamics at Mount St. Helens on 18 May 1980," *Nature* 291 (1981): 568–70.

11. Ibid.

12. Ibid.

13. The following assumptions were used for the calculations. Each F-1 engine has a power per unit area of 6.45 MW per square inch, or 929 MW per square foot (ft^2). The area of the exit of an F-1 engine is 105 ft^2. Therefore, the power of an F-1 engine is 97,550 MW, and that of a Saturn V is 487,740 MW. The power per unit area of the Mount St. Helens lateral blast is assumed to have been 2,322 MW/ft^2. The area of the vent for the lateral blast is assumed to be 2.7 million ft^2. Therefore, the total power of the lateral blast was 6,270 million MW, 12,855 times that of a Saturn V. The Fukushima power plant was projected to be able to produce 4,500 MW. The thrust of a Saturn V rocket is 7.6 million pounds force; the thrust of Mount St. Helens was 7.4×10^{11} pounds force. Numbers cited in the text have been rounded off.

14. D. E. Ogden, K. H. Wohletz, G. A. Glatzmaier, and E. E. Brodsky, "Numerical Simulations of Volcanic Jets: Importance of Vent Overpressure," *Journal of Geophysical Research* 113 (2008): B02204; H. Pinkerton, L. Wilson, and R. Macdonald, "The Transport and Eruption of Magma from Volcanoes: A Review," *Contemporary Physics* 43, no. 3 (2002): 197–210.

15. This information comes from the numerous papers in C. G. Newhall and R. S. Punongbayan, eds., *Fire and Mud: Eruptions and Lahars of Mount Pinatubo, Philippines* (Quezon City: Philippine Institute of Volcanology and Seismology, 1996), http://pubs.usgs.gov/pinatubo/prelim.html. Particularly

useful is E. W. Wolfe and R. P. Hoblitt, "Overview of the Eruptions," http://pubs.usgs.gov/pinatubo/wolfe.

16. P. Chakraborty, G. Gioia, and S. W. Kieffer, "Volcanic Mesocyclones," *Nature* 458 (2009): 497–500.

17. All material relating volcanic eruptions to mesocyclones is from Chakraborty, Gioia, and Kieffer, "Volcanic Mesocyclones."

18. Ibid.

19. C. G. Newhall and S. Self, "The Volcanic Explosivity Index (VEI): An Estimate of Explosive Magnitude for Historical Volcanism," *Journal of Geophysical Research* 87, no. C2 (1982): 1231–38.

20. US Geological Survey, "Report: Eruptions of Mount St. Helens: Past, Present, and Future," accessed February 18, 2011, http://vulcan.wr.usgs.gov/Volcanoes/MSH/Publications/MSHPPF/MSH_past_present_future.html.

21. For the real, not Hollywood, science of Yellowstone and its eruptions, see R. B. Smith and J. L. Siegel, *Windows into the Earth: The Geologic Story of Yellowstone and Grand Teton National Parks* (Oxford: Oxford University Press, 2000).

22. R. Evans, "Blast from the Past," *Smithsonian Magazine*, July 2002.

23. W. J. Broad, "It Swallowed a Civilization," *New York Times*, October 21, 2003, http://www.nytimes.com/2003/10/21/science/earth/21VOLC.html.

24. M. R. Rampino and S. Self, "Climate-Volcanism Feedback and the Toba Eruption of ~70,000 Years Ago," *Quaternary Research* 40 (1993): 269–80; M. R. Rampino and S. Self, "Volcanic Winter and Accelerated Glaciation following the Toba Super-eruption," *Nature* 359 (1992): 50–52.

25. S. H. Ambrose, "Late Pleistocene Human Population Bottlenecks, Volcanic Winter, and Differentiation of Modern Humans," *Journal of Human Evolution* 34 (1998): 623–51.

Chapter 6 THE POWER OF WATER: TSUNAMIS

1. This is not the only giant wave that has struck Lituya Bay. For more detail, see D. J. Miller, *Giant Waves in Lituya Bay, Alaska*, Geological Survey Professional Paper 354-C (Washington, DC: US Government Printing Office, 1960), 51–86.

2. In the past, meteorite impacts have generated huge tsunamis, but such

impacts do not appear to be a likely cause at the present time, because we no longer have large meteorites on a collision course with the Earth. A small meteorite hitting in the wrong place could still cause significant local damage, however.

3. C. L. Mader and M. L. Gittings, "Modeling the 1958 Lituya Bay Mega-tsunami," *Science of Tsunami Hazards* 20 (2002): 242.

4. The height was reconstructed from scoured trees and soil disrupted by the tsunami as it passed.

5. Delaware (451 feet), District of Columbia (409 feet), Florida (345 feet), Illinois (956 feet), Indiana (937 feet), Iowa (1,190 feet), Louisana (543 feet), Michigan (1,408 feet), Mississippi (807 feet), Missouri (1,542 feet), Ohio (1,095 feet), Rhode Island (812 feet), and Wisconsin (1,372 feet).

6. G. M. McMurtry, G. J. Fryer, D. R. Tappin, I. P. Wilkinson, M. Williams, J. Fietzke, D. Garbe-Schoenberg, and P. Watts, "Megatsunami Deposits on Kohala Volcano, Hawaii, from Flank Collapse of Mauna Loa," *Geology* 32 (2004): 741–44.

7. An excellent general source of information on tsunamis is E. Bryant, *Tsunami: The Underrated Hazard* (Cambridge: Cambridge University Press, 2001).

8. "The 26 December 2004 Indian Ocean Tsunami: Initial Findings from Sumatra," *USGS*, accessed July 14, 2012, http://walrus.wr.usgs.gov/tsunami/sumatra05/heights.html. And more recently, H. Gibbons and G. Gelfen-baum, "Astonishing Wave Heights among the Findings of an International Tsunami Survey Team on Sumatra," *Sound Waves*, March 2005, http://soundwaves.usgs.gov/2005/03.

9. E. R. Scidmore, "The Recent Earthquake Wave on the Coast of Japan," *National Geographic*, September 1896, http://ngm.nationalgeographic.com/1896/09/japan-tsunami/scidmore-text.

10. Surface tension is not important for most of the geological-scale processes discussed in this book and will be ignored hereafter.

11. Technically, these waves are controlled by gravity and inertia, where "inertia" means the resistance of an object to a change in its state of motion as described physically by Newton's first law. For small waves, surface tension between the water and air is a restoring force, but it is important only for very small waves, much smaller than we will be discussing in this book.

12. G. G. Stokes, "On the Steady Motion of Incompressible Fluids," *Transactions of the Cambridge Philosophical Society* 7 (1842): 439–54.

13. A review of the procedures for processing data can be found in A. Suppasri, F. Imamura, and S. Koshimura, "Tsunamigenic Ratio of the Pacific Ocean Earthquakes and a Proposal for a Tsunami Index," *Natural Hazards Earth System Science* 12 (2012): 175–85.

14. T. Fujiwara, S. Kodaira, T. No, Y. Kaiho, N. Takahashi, and Y. Kaneda, "The 2011 Tohoku-Oki Earthquakes: Displacement Reaching the Trench Axis," *Science* 334 (2011): 1240. Also in M. Fischetti, "Fukushima Earthquake Moved Seafloor Half a Football Field," *Scientific American*, December 1, 2011, http://www.scientificamerican.com/article.cfm?id=japan-earthquake-moves-seafloor.

15. See note 1 in Chapter 3.

16. G. R. Gisler, "Tsunami Simulations," *Annual Reviews of Fluid Mechanics* 40 (2008): 71–90.

17. Ibid.

18. Y. Fujii and K. Satake, "Off Tohoku-Pacific Tsunami on March 11, 2011," accessed July 14, 2011, http://iisee.kenken.go.jp/staff/fujii/OffTohokuPacific2011/tsunami.html. This document was posted on March 12, 2011, one day after the Tohoku earthquake.

19. The dimensions assumed for this calculation are as follows: For a wind-driven wave: height 50 feet, breadth 10,000 feet (about 2 miles), wavelength 500 feet, giving a volume of 250 million cubic feet. For a tsunami: height on the open ocean 10 feet, breadth 100 miles, wavelength 500 miles (2,640,000 feet), giving a volume of 14 trillion cubic feet.

20. The tsunami power calculation is at http://plus.maths.org/content/tsunami-1. *Plus* magazine is dedicated to making mathematics a living subject.

21. In fact, because of the geometry of the offshore region around Sri Lanka, the waves broke offshore, and a substantial fraction of the power was dissipated in the breaking waves before they hit landfall. If this had not been the case, the disaster would have been much worse.

22. T. Maeda, T. Furumura, S. Sakai, and M. Shinohara, "Significant Tsunami Observed at the Ocean-Bottom Pressure Gauges during the 2011 [Earthquake] off the Pacific Coast of Tohoku Earthquake," *Earth, Planets and Space* 63 (2011): 803–8.

23. P. Hancocks, "Defiant Japanese Boat Captain Rode Out Tsunami," *CNN World*, April 3, 2011, http://www.cnn.com/2011/WORLD/asiapcf/04/03/japan.tsunami.captain/index.html?hpt=C2.

24. A video of a Japanese coast guard vessel, the *Matsushima*, riding over the

crest of two of the March 11, 2011, tsunami waves 3 miles out at sea shows the nature of the tsunami close to the coast, but still far enough out that the wave is not breaking. It can be found at "Footage of Japan Tsunami Waves at Sea Captured by Coast Guard," *Telegraph*, March 19, 2011, http://www .telegraph.co.uk/news/worldnews/asia/japan/8392419/Footage-of-Japan-tsunami-waves-at-sea-captured-by-Coast-Guard.html.

25. P. Neale, "The Krakatoa Eruption," in *Littells Living Age*, 5th series, vol. 51 (Boston: Littell and Co., 1885), 693–96. Digital copy available at http:// books.google.com.

26. K. Minoura, F. Imamura, D. Sugawara, Y. Kono, and T. Iwashita, "The 869 Jogan Tsunami Deposit and Recurrence Interval of Large-Scale Tsunami on the Pacific Coast of Northeast Japan," *Journal of Natural Disaster Science* 23, no. 2 (2001): 83–88.

27. Ibid. The reader will note that the 1896 Meiiji-Sanruku tsunami discussed earlier in this chapter is not included in this discussion. It was generated on a different fault segment north of the one that caused the three tsunamis under discussion.

28. P. R. Cummins, "The Potential for Giant Tsunamigenic Earthquakes in the Northern Bay of Bengal," *Nature* 449 (2007): 75–78.

Chapter 7 ROGUE WAVES, STORMY WEATHER

1. P. N. Joubert, "Some Remarks of the 1998 Sydney-Hobart Race," *Transactions of the Proceedings of the Royal Society of Victoria* 11, no. 2 (2006): i–x.

2. Strictly, "mad dog wave" is a Taiwanese fishermen's term for near-shore rogue waves, generalized here to include all rogue waves.

3. C. Kharif, E. Pelinovsky, and A. Slunyaev, *Rogue Waves in the Ocean*, Advances in Geophysical and Environmental Mechanics and Mathematics (New York: Springer, 2009), 20.

4. P. C. Liu and U. F. Pinho, "Freak Waves—More Frequent Than Rare!" *Annales Geophysicae* 22 (2004): 1839–42.

5. J. Sheldon, "Cruise Ship Hit by Rogue Wave, 2 Killed," *Manolith*, March 4, 2010, http://www.manolith.com/2010/03/04/cruise-ship-hit-by-rogue-wave-2-killed. An interesting video about the 100-foot wave that hit the oil tanker *Esso Languedoc*, and about the rogue waves that hit the oil platforms

Ocean Ranger and *Drovner*, can be found at http://www.youtube.com/watch?v=sCxr_XzyGO8&feature=related.

6. C. B. Smith, *Extreme Waves* (Washington, DC: Joseph Henry Press, 2006), 4.

7. Kharif, Pelinovsky, and Slunyaev, *Rogue Waves in the Ocean.*

8. Smith, *Extreme Waves.*

9. To provide a reference for defining rogue waves, oceanographers have introduced the concept of "significant wave height." The significant wave height is the average wave height of the one-third highest waves in a time period (typically taken as ten to thirty minutes). Surfers might find the following exercise useful—and sobering: Ignoring the small stuff, sit on a beach and make a list of the heights of all incoming waves for ten to thirty minutes. For example, 1 foot (1'), 2 feet (2'), 3', 5', 3', 4', and so on. Organize this list from biggest to smallest: 5', 4', 3', 3', 2', 1'. Keep the highest one-third of the values, 5' and 4'. Then take their average: 4.5'. This is the significant wave height. A rogue wave is defined as one whose height is two or more times the significant wave height—at least 9' in this example. In fact, rogue waves with a height more than four times the significant wave height have been documented. In this example, that would be an 18'-high wave. Surfers who are comfortable with 2' or 3' waves, perhaps an occasional 5' wave, but not with 9' or 18' waves need to be aware that such waves can appear at any time. This is apparently what happened when a wall of water collapsed on and killed the experienced big-wave surfer, Sion Milosky, at Half Moon Bay, California, in March 2011. After nearly an hour of relatively small swells (18'–20') for that day, a rogue wave "bomb" rolled Milosky to the bottom, where he was held down not only by this wave, but by a second wave as well, in what is known in surfing jargon as a "two-wave hold-down." He was found too late, twenty minutes later.

10. Some of the best discussion of this wave is in the informal literature. Much of the material related here is taken from the report by Sverre Haver titled "A Possible Freak Wave Event Measured at the Draupner Jacket January 1 1995" (http://www.ifremer.fr/web-com/stw2004/rw/fullpapers/walk_on_haver.pdf). See also E. Bitner-Gregersen, "Extreme Wave Crest and Sea State Duration," Appendix B4 in A. D. Jenkins et al., *Research Report No. 138* (Bergen, Norway: Norwegian Meteorological Institute, 2002), 97.

11. Deep and asymmetrical "holes" also exist, but for simplicity they are not discussed here.

12. R. W. Warwick, "Hurricane 'Luis,' the *Queen Elizabeth 2* and a Rogue

Wave," *Marine Observer* 66 (1996): 134. A more accessible account can be found at http://www.cruiseshipsinking.com/Damaged_By_Waves/Queen_Eliz abeth_2_High_Waves_September_11_1995.html, accessed February 13, 2013.

13. Although it would be of great interest to know the area of the ocean covered by the instruments during the three weeks of observations, I was not able to make that determination from the data publicly available.

14. B. Gaine, "Predicting Rogue Waves," *MIT Technology Review*, March 1, 2007, http://www.technologyreview.com/computing/18245/page2.

15. B. Baschek and J. Imai, "Rogue Wave Observations off the US West Coast," *Oceanography* 24, no. 2 (2011): 158–65.

16. H. J. Lugt, *Vortex Flow in Nature and Technology* (New York: Wiley, 1983); also Haver, "Possible Freak Wave Event." Lugt cites many figures and facts that apparently appeared first in E. N. Lorenz, *The Nature and Theory of the General Circulation of the Atmosphere* (Geneva, Switzerland: World Meteorological Organization, 1967).

17. Specifically, Hadley used linear momentum instead of angular momentum around Earth's rotational axis.

18. G. G. Coriolis, "Mémoire sur les équations du mouvement relatif des systèmes de corps," *Journal de l'École Polytechnique* 15 (1835): 142–54.

19. W. Ferrel, "An Essay on the Winds and Currents of the Ocean," *Nashville Journal of Medicine and Surgery* 11 (1856): 287–301.

20. The visualization entitled "Perpetual Ocean," at http://svs.gsfc.nasa.gov/ vis/a010000/a010800/a010841 shows two years (2005–07) of beautiful documentation of the ocean currents. It was created by H. Mitchell and G. Shirah and produced by the Scientific Visualization Studio at NASA's Goddard Space Flight Center.

21. Smith, *Extreme Waves*, 187.

22. There is an excellent quantitative discussion (that is, mathematical, but explained clearly with good graphs) of wave development and significant wave heights on the website OceanWorld run by the Jason Education Project at Texas A&M University: http://oceanworld.tamu.edu/resources/ocng_ textbook/chapter16/chapter16_04.htm (accessed July 19, 2012).

23. Smith, *Extreme Waves*, 188.

24. Interestingly, Dias first named this place the "Cape of Storms" (*Cabo das Tormentas*), but the name was changed to the "Cape of Good Hope" by King John II of Portugal because it opened trade routes to the east that bypassed the middlemen in the Middle East.

25. "Agulhas Current," *Weathernews.com*, September 2009, http://weath ernews.com/TFMS/topics/wtopics/2007/pdf/20070901.pdf.

26. F. Baronio, A. Degasperis, M. Conforti, and S. Wabnitz, "Solutions of the Vector Nonlinear Schrödinger Equations: Evidence for Deterministic Rogue Waves," *Physical Review Letters* 109 (2012): 044102–4. For technical papers related to these topics, see *European Physical Journal Special Topics* 185, issue 1 (July 2010).

Chapter 8 RIVERS IN THE SKY

1. R. Chandler, "Red Wind," *Dime Detective Magazine*, January 1938.

2. T. Finnigan, "Hydraulic Analysis of Outflow Winds in Howe Sound, British Columbia" (master's thesis, University of British Columbia, 1991). Finnigan reviews earlier work, so detailed citations are not given here.

3. "The Big Chill Heads South," *AP News Archive*, February 1, 1989, http:// www.apnewsarchive.com/1989/The-Big-Chill-Heads-South/id-38d2fe33178 58dbd49289094d09911c7; P. Anderson, "Alaska Blaster: Cold Wave Brings Big Chill to Ever-Widening Area," *AP News Archive*, February 2, 1989, http://www.apnewsarchive.com/1989/Alaska-Blaster-Cold-Wave-Brings-Big-Chill-to-Ever-Widening-Area/id-c9ee9aa7c57b045c3ee98bffd4dfe3d9.

4. T. R. Reed, "Gap Winds in the Strait of Juan de Fuca," *Monthly Weather Review* 59, no. 10 (1931): 373–76.

5. A. G. Sulzberger and B. Stelter, "A Rush to Protect Patients, Then Bloody Chaos," *New York Times*, May 23, 2011, http://www.nytimes .com/2011/05/24/us/24tornado.html?pagewanted=1&_r=1.

6. Death toll as of June 13, 2011.

7. E. H. Shackleton and T. W. E. David, *The Heart of the Antarctic*, popular ed. (Philadelphia: Lippincott, 1914), 72–74.

8. S. Kieffer, "The 1983 Hydraulic Jump in Crystal Rapid: Implications for River-Running and Geomorphic Evolution in the Grand Canyon," *Journal of Geology* 93 (1985): 385–406.

9. G. Gioia, P. Chakraborty, S. F. Gary, C. Z. Zamalloa, and R. D. Keane, "Residence Time of Buoyant Objects in Drowning Machines," *Proceedings of the National Academy of Sciences of the USA* 108 (2011): 6361–63.

10. Weather and climate are distinguished by how much time is involved. "Weather" refers to conditions in the atmosphere over fairly short periods of

time: days, weeks, or months. "Climate" refers to conditions averaged over a longer time.

11. No records exist that would allow us to determine how the winds that Shackleton described arose; there may have been a regional storm in the area, but we can ignore that detail here.

12. See, for example, L. J. Cooke, M. S. Rose, and W. J. Becker, "Chinook Winds and Migraine Headache," *Neurology* 54 (2000): 302.

13. The term "rotor" comes from the technical description of the geometry of the waves.

14. J. Kuettner and R. F. Hertenstein, "Observations of Mountain-Induced Rotors and Related Hypotheses: A Review" (paper presented at the 10th Conference on Mountain Meteorology, sponsored by the American Meteorological Society, Park City, Utah, 2002), 326–29.

15. A summary of this effort can be found at *Wikipedia*, "Perlan Project," accessed July 20, 2012, http://en.wikipedia.org/wiki/Perlan_Project.

16. M. G. Wurtele, "Meteorological Conditions Surrounding the Paradise Airline Crash of March 1964," *Journal of Applied Meteorology* 9 (1970): 787–95.

17. *Wikipedia*, "BOAC Flight 911," accessed February 13, 2013, http://en.wikipedia.org/wiki/BOAC_Flight_911.

18. Reed, "Gap Winds."

19. *American Meteorological Society Glossary of Meteorology*, "gap wind" entry, accessed February 13, 2013, http://glossary.ametsoc.org/wiki/Gap_wind.

20. Summarized in Finnigan, "Hydraulic Analysis."

21. Summarized in F. K. Ball, "The Theory of Strong Katabatic Winds," *Australian Journal of Physics* 9 (1956): 373–86.

22. G. Goebel, "Balloon Bombs against the US," accessed July 20, 2012, http://www.axishistory.com/index.php?id=932.

23. In general, nor'easters start as a storm in the Gulf of Mexico, but the polar jet stream plays a prominent role in drawing them northward.

24. "February Blizzard Strikes U.S. Northeast," Earth Observatory, February 13, 2013, http://earthobservatory.nasa.gov/IOTD/view.php?id=80412.

25. "Post-landfall Loss Estimates for Superstorm Sandy Released," *EQECAT*, November 1, 2012, http://www.eqecat.com/catwatch/post-landfall-loss-estimates-superstorm-sandy-released-2012-11-01.

26. E. S. Blake, C. W. Landsea, and E. J. Gibney, "The Deadliest, Costliest, and Most Intense United States Tropical Cyclones from 1851 to 2010 (and

Other Frequently Requested Hurricane Facts)," NOAA Technical Memorandum NWS NHC-6, (Miami, FL: National Weather Service, 2011), http://www.nhc.noaa.gov/pdf/nws-nhc-6.pdf.

27. The National Oceanic and Atmospheric Administration's summary of the events of spring 2011 is at NOAA, National Climatic Data Center, "Spring 2011 U.S. Climate Extremes," accessed February 13, 2013, http://www.ncdc.noaa.gov/special-reports/2011-spring-extremes/index.php; and the detailed report on Joplin is at NOAA, National Climatic Data Center, "State of the Climate: Tornadoes, May 2011," accessed February 13, 2013, http://www.ncdc.noaa.gov/sotc/tornadoes/2011/5.

28. The following material is from M. Levitan, "May 22, 2011 Joplin, MO Tornado Study: Draft Study Plan and Research Overview," accessed April 25, 2013, http://www.nist.gov/el/disasterstudies/ncst/upload/NCSTACJoplin110411.pdf.

Chapter 9 WATER, WATER EVERYWHERE . . . OR NOT A DROP TO DRINK

1. A. Coopes, "Australian Floods Expected to Peak at Rockhampton," *Sydney Morning Herald*, January 4, 2011, http://news.smh.com.au/breaking-news-world/australian-floods-expected-to-peak-at-rockhampton-20110104-19eya.html.

2. M. Wisniewski, "Midwest Braces for Massive Winter Blizzard," *Reuters*, January 31, 2011, http://www.reuters.com/article/2011/01/31/us-weather-midwest-storm-idUSTRE70U4NP20110131.

3. "Chicago Blizzard: Massive Winter Storm Hits Chicago," *Huffington Post*, January 31, 2011 (updated May 25, 2011), http://www.huffingtonpost.com/2011/01/31/chicago-blizzard-massive-storm_n_816673.html.

4. *Wikipedia*, "2011 Souris River Flood," accessed February 13, 2013, http://en.wikipedia.org/wiki/2011_Souris_River_flood.

5. NOAA, National Climatic Data Center, "State of the Climate: National Overview, July 2011," accessed August 11, 2012, http://www.ncdc.noaa.gov/sotc/national/2011/7.

6. D. Huber, "The 2011 Texas Drought in a Historical Context," *Center for Climate and Energy Solutions*, August 26, 2011, http://www.c2es.org/blog/huberd/2011-texas-drought-historical-context.

7. US Drought Monitor Report, National Drought Mitigation Center, August 7, 2012, available under "Tabular Statistics" at http://droughtmonitor.unl.edu.

8. Water in the ground flows toward these same places, though in a more complex way than surface water does.

9. Along the East Coast are three major basins: the Great Lakes–St. Lawrence River basin, the North Atlantic basin that extends from southern Virginia up along the coast to Canada, and the South Atlantic–Gulf of Mexico watershed in the southern states. Along the West Coast are the Columbia River–Northwestern basin, California east of the Sierra Nevada, the Great Basin of Nevada-Utah-California, the Colorado River basin in the Four Corners area, the Rio Grande river basin, and the Gulf Coast. By far the largest watershed in the US is the Mississippi-Missouri basin. Watersheds transcend local, state, regional, and national boundaries. They can have very different characteristics depending on the regional climate.

10. P. M. Cox, R. A. Betts, M. Collins, P. P. Harris, C. Huntingford, and C. D. Jones, "Amazonian Forest Dieback under Climate-Carbon Cycle Projections for the 21st Century," *Theoretical and Applied Climatology* 78 (2004): 137–56.

11. Quoted from remarks to the Australian State Governors Sydney Futures Forum by Anne Davies, reported in the *Sydney Morning Herald*, May 19, 2004. For a more general discussion, see T. Flannery, *The Weather Makers: How Man Is Changing the Climate and What It Means for Life on Earth* (New York: Grove Press, 2005).

12. This phrase is from the following excellent fact sheet on El Niños and La Niñas: B. Hensen and K. E. Trenberth, "Children of the Tropics: El Niño and La Niña," February 1998 (updated October 2001), http://www.ucar.edu/communications/factsheets/elnino.

13. N. Nicholls, "El Niño—Of Droughts and Flooding Rains," accessed February 13, 2013, http://www.abc.net.au/science/slab/elnino/story.htm; R. G. Kimber, "Australian Aboriginals' Perceptions of Their Desert Homelands (Part 1)," *Arid Lands Newsletter* no. 50, November/December 2001, http://ag.arizona.edu/oals/ALN/aln50/kimberpart1.html.

14. Nicholls, "El Niño."

15. See, for example, NOAA, National Climatic Data Center, "ENSO Technical Discussion," accessed February 13, 2013, http://www.ncdc.noaa.gov/teleconnections/enso/enso-tech.php.

16. NOAA, "What Is La Niña?," accessed February 13, 2013, http://www
.pmel.noaa.gov/tao/elnino/la-niña-story.html.

17. An excellent reference is National Weather Service Climate Prediction
Center, "Frequently Asked Questions about El Niño and La Niña, accessed
February 13, 2013, http://www.cpc.ncep.noaa.gov/products/analysis_moni
toring/ensostuff/ensofaq.shtml#DROUGHTETC.

18. See, for example, the explanation and graphics in R. Knabb, "Birth of
a Hurricane: Where They Come From," *The Weather Channel*, accessed
February 13, 2013, http://www.weather.com/outlook/weather-news/news/
articles/hurricanes-where-do-they-come-from_2011-07-22?page=3.

19. An excellent interactive time-lapse movie of storms nucleating and moving out from Africa can be found at http://www.weather.com/weather/
map/interactive/Ouagadougou+Burkina%20Faso+UVXX0001?zoom=4,
accessed February 13, 2013. Zooming in and out on this map allows tracking
of storms globally. Loading may take some time.

20. C. Dolce, "Hurricane Irene's Alleyway," *The Weather Channel*, August
24, 2011, accessed August 15, 2012, http://www.weather.com/weather/
hurricanecentral/article/irenes-alleyway-north_2011-08-23.

21. Hurricane Sandy occurred just when the La Niña of 2011 had been
declared over and the weather was transitioning into either a weak El Niño
or a neutral year that would be neither La Niña nor El Niño.

22. E. N. Lorenz, *The Nature and Theory of the General Circulation of the
Atmosphere* (Geneva, Switzerland: World Meteorological Organization, 1967).

23. Q. Schiermeier, "Extreme Measures," *Nature* 477 (2011): 148–49.

24. R. Dole, M. Hoerling, J. Perlwitz, J. Eischeid, P. Pegion, T. Zhang, X.-W.
Quan, T. Xu, and D. Murray, "Was There a Basis for Anticipating the 2010
Russian Heat Wave?," *Geophysical Research Letters* 38 (2011): L06702,
doi:10.1029/2010GL046582.

25. T. C. Peterson, P. A. Stott, and S. Herring, eds., "Explaining Extreme
Events of 2011 from a Climate Perspective," *Bulletin of the American Meteorological Society* 93 (2012): 1041–67.

26. G. Van Oldenborgh, G., A. van Urk, and M. R. Allen, "The Absence of
a Role of Climate Change in the 2011 Thailand Floods," in Peterson, Stott,
and Herring, "Explaining Extreme Events of 2011."

27. C. C. Funk, "Exceptional Warming in the Western Pacific-Indian Ocean
Warm Pools Has Contributed to More Frequent Droughts in Eastern Africa,"
in Peterson, Stott, and Herring, "Explaining Extreme Events of 2011."

28. The authors used the drought of 2008 as a proxy for the 2011 drought because models were not available for the latter.

29. D. E. Rupp, P. W. Mote, N. Massey, C. J. Rye, R. Jones, and M. R. Allen, "Did Human Influence on Climate Make the 2011 Texas Drought More Probable?," in Peterson, Stott, and Herring, "Explaining Extreme Events of 2011."

30. According to Debarati Guha Sapir, director of the Centre for Research on the Epidemiology of Disasters at the Catholic University of Louvain in Belgium, quoted in L. Schlein, "2011 Costliest Year in History for Catastrophes," *Voice of America*, January 17, 2012, http://www.voanews.com/content/ article-2011-costliest-year-in-history-for-catastrophes-137585693/159469 .html.

31. M. Lagi, K. Z. Bertrand, and Y. Bar-Yam, "The Food Crises and Political Instability in North Africa and the Middle East," *Social Science Research Network*, August 15, 2011, http://ssrn.com/abstract=1910031.

32. P. Krugman, "Droughts, Floods and Food," *New York Times*, February 6, 2011, http://www.nytimes.com/2011/02/07/opinion/07krugman.html.

33. Ibid.

Chapter 10 EARTH AND US

1. Some of this account and all of the quotes are from S. S. Hall, "Scientists on Trial: At Fault?" *Nature* 477 (2011): 265–69.

2. D. Petley, "Attempts to Predict Earthquakes May Do More Harm Than Good," *Guardian*, May 3, 2012, http://www.guardian.co.uk/science/ blog/2012/may/30/attempts-predict-earthquakes-harm-good.

3. IAVCEI Subcommittee for Crisis Protocols, "Professional Conduct of Scientists during Volcanic Crises," *Bulletin of Volcanology* 60 (1999): 323–34.

4. "Japan Readies for Reopening of Nuclear Reactors amid Safety Concerns," *Guardian*, June 8, 2012, http://www.guardian.co.uk/world/2012/jun/08/japan -reopen-nuclear-reactors-safety.

5. Parts of the discussion in this chapter were presented in S. W. Kieffer, P. Barton, W. Chesworth, A. R. Palmer, P. Reitan, and E. Zen, "Megascale Processes: Natural Disasters and Human Behavior," in *Preservation of*

Random Megascale Events on Mars and Earth: Influence on Geologic History, Geological Society of America Special Papers 453, ed. M. G. Chapman and L. P. Keszthelyi (Boulder, CO: Geological Society of America, 2009), 77–86.

6. Now referred to as "Bureau approval."

INDEX

Page numbers in italics refer to illustrations.